The
Shearers

PENGUIN

UK | USA | Canada | Ireland | Australia

India | New Zealand | South Africa | China

Penguin is an imprint of the Penguin Random House group of companies, whose
addresses can be found at global.penguinrandomhouse.com.

First published by Penguin Random House New Zealand, 2019

1 3 5 7 9 10 8 6 4 2

Text © Ruth Entwistle Low, 2019

Photography © Mark Low, 2019

The moral right of the author has been asserted.

Cover and text design by Megan van Staden © Penguin Random House New Zealand

Photography by Mark Low unless otherwise credited

Front cover photograph: Charles O'Neill, ganger for Peter Lyon Shearing, Mt Nicholas
Station, Lake Wakatipu

Back cover photograph, top: Bringing in the merinos for shearing at Mt Nicholas
Station, Lake Wakatipu; bottom: Brian Kerr setting up his handpiece for the next run,
Waerenga, Waikato

Prepress by Image Centre Group

Printed and bound in China by RR Donnelley

A catalogue record for this book is available from the National Library of
New Zealand.

ISBN 978-0-14-377116-6

eISBN 978-0-14-377117-3

penguin.co.nz

The Shearers

New Zealand Legends

Ruth Entwistle Low
Photography by Mark Low

PENGUIN BOOKS

Shearers work in some of the most remote, beautiful country in New Zealand. Quail Flat woolshed, Muzzle Station.

To Stewy, my writing companion
and more importantly, my dad.

Contents

Introduction

Driving up to any one of the myriad woolsheds around the country at shearing time, you'll invariably find a host of vehicles parked outside: the owner's ute, with all manner of farming paraphernalia on its flatdeck; the contractor's ute, maybe with the tailgate open revealing a spare grinder waiting to be brought into the shed if needed; and, of course, the ubiquitous Toyota HiAce or Ford Transit vans with the contractor's details splashed across the sides. Emerging from your vehicle you instantly become aware of activity. If the big sliding door of the shed is closed, you hear a low but incessant thrum; if the door is ajar the sound is more assaulting: loud music merges with the drone of shearing machines. There may also be the sound of cloven-hoofed ovines running over wooden gratings while harried from holding pen to catching pen, ready for shearing. Adding to the intensity of sound and action is their bleating and the occasional sharp bark from a working dog.

Care is called for as you walk up muddy steps to the landing of the shed; chicken wire nailed tightly over each step warns of the potential to slip. Then, standing at the entrance to the shed, you look in, immediately feeling like a rank outsider as you take stock of all that is unfolding. Depending on the size of the shed there could be anything from a handful of people to the equivalent of a rugby team plus subs working inside. They are all in a constant state of motion, but you sense they know that 'foreigners' are watching. You are met with equal parts curiosity and suspicion. It is very easy for an urbanite to feel intimidated.

How does one absorb and make sense of it all? You are slow to enter for fear of upsetting the carefully choreographed routine. There is no mistaking the shearers on the 'board' – their handpieces moving snakelike across the sheep's skin, the machine power-cord pulled at the beginning and end of each decloaking (the frequency of the action denoting the shearer's experience level). But as to the other cohort, it takes time to understand the pattern of their movement. There are those sweeping wool away from shearers with paddle-like brooms, and those coming to pick up each fleece and throw it on a central table where others gather round, grabbing rapidly at the fleece and removing anything detracting from its quality. One of those around the table folds up the fleece and passes it on to a circular revolving table where someone, looking official, examines each fleece and makes a decision as to what bin, of which there are several, it will be placed into. There's a person who comes over, takes great armfuls of wool out of a bin and loads it into a press, where it is eventually compressed into a bale. Somewhere close by, a pile of wool bales is building up, marking the number of sheep that have already passed over the board. All the time the machines and music batter your ears, and there's the odd bit of chatter and laughter. In the corner of the shed the cocky and contractor may be chatting, but their eyes sweep the scene, keenly observing the team in motion. Somehow all these individuals work in unison to ensure that the cocky's sheep are shorn and the wool processed as quickly and with as little fuss as possible.

When brave enough to enter you shake hands with the cocky and they talk to you about their sheep, their breed and what they are aiming to produce. There are the fine wool growers, ensuring the fleeces of their merinos are treated with considerable respect, as the fleece is a valuable commodity. Then there are the cross-breds, which produce coarse wool; the farmer's eye then is perhaps not quite so sharply attentive on the wool. In today's market coarse wool generates little income, its value having gradually declined to a point where, for many, shearing has become about maintaining animal health rather than seeing a substantial wool cheque augment their income. Fortunately for them the lamb market is buoyant, so, with an eye to meat, they talk of changing farming practices to optimise profit.

Next, you catch up with the contractor, who explains the roles of those in the shed. Of course, the shearers need no explanation; but there is conversation around who are the best in the team. The fastest shearer – or, in old parlance, the ringer – is in the middle stand, close to the table to ensure the wool flows easily to the woolhandlers. With another sheep down the porthole he straightens, clicks his tally counter, wipes his brow on the towel hanging on a string across the frame of the pen door and swings the catching door open; selecting his next target, he grabs it firmly under its chin, tips it up and drags it out, positioning it nicely, the animal relaxed. Then the shearer picks up his handpiece, pulls the cord and starts down the animal's brisket or chest. The movement is smooth and controlled, the sweat beading on his brow the clear evidence of exertion.

Your head turns to someone at the last stand and you begin to realise that perhaps it is not as easy as it looks. A young shearer struggles to position his ewe correctly and it is kicking out. The contractor slips over to give him a hand, signalling to move his foot around into the sheep more. Once positioned properly, the contractor pulls the cord for the struggler and stands patiently watching, offering words of encouragement and further advice when needed. Once the sheep is down the porthole the contractor grabs a ewe out of the pen, takes the handpiece, pulls the cord and starts to peel the wool off; the struggler seizes the opportunity to stretch his back, then stands by, taking in every move, studying the placement of his boss's feet.

Once the lesson is over the contractor is back beside you explaining the ins and outs of the woolhandlers', pressers' and classers' jobs. Woolhandling terms fly out at a rate of knots – fribs, dags, crutching, skirting, VM (vegetable matter), lines, cotty ... Meanwhile, you watch each fleece being gathered, walked to the table and unfurled with a fluid throw; there is a moment to see if it settles perfectly on the table or lies draping over the edge. Again, you are aware that the expert hands make it look easy. It's a mesmerising spectacle.

There is the occasional cry of 'sheepo', the universal call for the presser or young lad out in the pens to fill the shearers' catching pen. The shearer is down to their last sheep and nothing must interrupt their flow. The pen is quickly filled, with lifting of gates, shoving of sheep and bleating all part of the equation. And as you take the time to watch and glean, the necessity for teamwork becomes evident. One member of a shearing gang cannot function without the other: each has a distinct and necessary role to play. But your eyes can't help but be drawn back in admiration to the shearer. As they click their counter, wipe their face and launch in to grab yet another sheep, you wonder what it is that drives them.

I have been that person climbing the stairs to the landing of a woolshed, and I was that person standing nervously at its entrance and watching until deeming it appropriate to enter. An urbanite with no farming experience, I probably looked much like a possum caught in headlights, or a foreigner in a strange land. I have sought the citizens of this 'land' – those who defleece the country's twenty-seven million-odd sheep – to enlighten me on how and why they do the work that they do.

Obviously, with hundreds of shearers working in the country at any given time I was not about to interview them all; choices had to be made. My door into this world was graciously opened firstly by the farmers and station owners I approached, each chosen for regional variance. They agreed to be interviewed, and their interviews undergirded or contextualised those that I was to have later with the shearers they recommended. Then, of course, because of my own penchant for interviewing 'just one more', there were a few extra shearers I nabbed along the way. Hopefully, all those many shearers not interviewed will feel very much a part of this story, and that the profession is honoured.

I have covered many miles and walked into quite a few woolsheds in pursuit of the shearers, who related their experiences and enlightened me on the skills and culture of their craft, and on what it means to be part of the shearing fraternity. Their interviews are the basis for this book, and their voices resound through these pages. Their stories all offer a window into the world of shearing, both past and present.

The book is divided into four thematic parts: The Shearers; Their Work; Their World; and lastly Their Guts and Glory. Vignettes of the interviewees then fit within these four broad themes. Part one contains a cross-section of some of the people you are likely to meet in the shed – the older, the younger, the Māori, the Pākehā, the women, the professional shed shearer, the competitive shearer, a blade shearer, and those who shore for a short time. Part two covers the day-to-day work of a shearer, revealing what a newbie will go through in order to become a professional, the mindset of a top shed shearer, the need to use your brain over mere brawn, the development of young blood, and the work of a contractor. Underlying the accounts is the incredible work ethic of the shearers. Part three explores the culture of shearing life: the camaraderie, the stay-outs, the sense of whānau, the humour and the 'work hard, play hard' attitude, as well as touching on how some of that culture is changing to encourage a stronger professional image. Part four introduces a selection of those who have done well in the competitive realm; while not all have reached the pinnacles of shearing glory, their stories emphasise the satisfaction gained from competing, as well as underpinning the significance of a sport that positively promotes the shearing industry. Much of the story is told directly by the shearers themselves and is based on their own accounts. Where they are directly quoted, for clarity, there has been minor editing, but care has been taken to ensure the meaning has not been altered.

I have nothing but admiration for shearers; theirs is one of the most physically demanding jobs in the world. As was said to me on several occasions, it's the only occupation where you take a sweat towel to work. For those who have never entered a shearing shed and seen the intensely physical world of the shearers in action, my hope is that by reading this book you will gain that same sense of admiration for the men and women who shear the country's sheep.

TOP Each run the ganger reads the shearer's counter and records the number of sheep shorn by them in the tally book. With a shearer paid at a piece rate they keep a close eye on their tallies.

BOTTOM The modern handpiece is a completely different beast compared to the old 'bricks' of the early days.

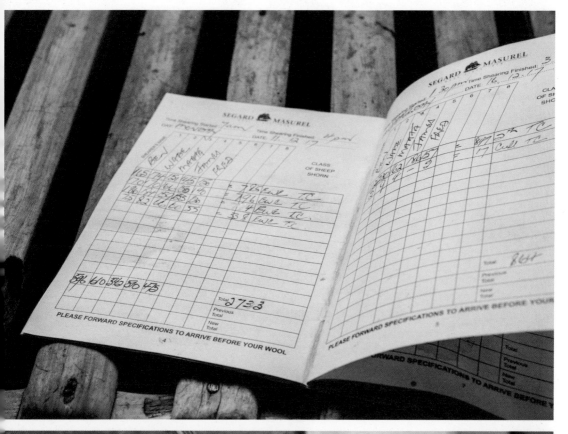

Shearing sheds around the country are adorned with the names of those who have worked there. While today a Sharpie is more likely to be the medium for the graffiti, in the past the stencils and ink used to brand the wool bales were commonly used. Brewers' farm, Taranaki.

1

The Shearers

This is probably the only country in the world where the national sheep tally makes headline news. For so long, New Zealand's wealth was quite literally made off the sheep's back, so it is perhaps unsurprising that the size of the national flock inevitably became a marker of the country's well-being.

The size of that flock has been declining for some decades, and it will not shock you to learn that today there are only six sheep per head of human population, compared to a peak of twenty in 1982; nevertheless, somewhere deep in the New Zealander's psyche is an awareness of the abiding importance of sheep to our economy.

Today, sheep meat exports are more economically important than wool. There is the wonderful success story of our fine wool industry, but it meets a niche market; the bulk of the wool grown in New Zealand is coarse wool, and for now its value languishes. No matter what its value, for the well-being of the animal the wool must come off – and the only efficient way to denude the flock is to shear them.

The shearing industry today is a far cry from its infancy in the colonial period, when a hodgepodge of itinerant male workers, most with little knowledge or experience, shore the burgeoning flocks. Over time, what has developed is a professional body of individuals with skills far more refined than the early shearers; the increase in individual shearing tallies gives a clear indication of that.

Traditionally it has been men who have shorn. In fact, aside from the Māori whānau gangs that quickly dominated the East Coast of the North Island, women were not part of woolshed life in some parts right up until the 1970s. The all-male sheds were colloquially known as 'ball-bearing gangs'. Reflecting the sensibilities of the day, shearers called out 'ducks on the pond' or 'ninety-nine' when women were about to enter the shed, to curb the use of any bad language.

Māori men have almost from the outset made up a large proportion of the workforce and quickly began to dominate the shearing boards. On the many big sheep stations on the East Coast it was solely Māori who shore. They forged a tradition of travelling from station to station over the summer shearing season as whānau gangs developed. This is believed to be the precursor to the contracting system now so well ingrained in the industry today. A fair few cockies' sons took up the handpiece, too, seeing shearing as a means to earn and save towards farm ownership. Many of them shore in the small two- and three-stand sheds around their local communities. As well, there were many from a farming background who would pick up a handpiece and shear for a time as necessity dictated.

When push came to shove, such as with wartime labour shortages, women stepped up. Over the years more women have taken up the handpiece and made their mark in the industry, although they are still in the minority. With the high level of testosterone in the shed, and the tradition of a strong macho culture, it's fair to say that not all have had an easy ride staking their claim. But as long as they prove themselves, just as any shearer has to, they are largely accepted. A common theme reiterated throughout the interviews is that what women may lack in brawn they make up for in the quality of their shear.

Whether female or male, Māori or Pākehā, or an international over here to hone their skills, what matters most of all is character. To take on shearing professionally requires guts, mental fortitude and a whole pile of determination, especially when a gang's at one of the big sheds and they're facing thousands of sheep. That's when the character of the shearer counts – when all they can do is 'keep walking into them, boy', as Reg Benjamin was instructed when a fledgling shearer.

Blade shearers in New Zealand are almost an endangered breed these days; they are found now only in sheds in pockets of the high country. Given, though, that the shearing industry was founded on them, it seems appropriate to acknowledge their significance and allow one of the 'knights of the blades', character and legend Peter Casserly, to be heard first. The Master Blade Shearer, one-time world champion and current world record holder has many stories of his experiences working the blades throughout Canterbury and Otago. His words reveal his love for an industry that he has been part of for decades.

Peter Casserly – The caveman

'They call us the cavemen cos we're still using stones to sharpen our gear, and we call them the bog iron men. You know – they've got the iron in the hand.'

Peter is not backward in coming forward, and, while supping a beer in his man-cave in Ōmarama with shearing memorabilia close at hand, he happily expounds his views on anything from who the best shearers are today to how he believes blade shearers are treated as second-class citizens by the shearing fraternity. 'They forget that we were there first . . . I've got to tell John Fagan about that over a bottle of rum one night.' With the same enthusiasm that he takes on any controversial topic, Peter launches into recollections.

Shearing was an unlikely occupation for a young lad growing up in the mining community of Dobson, just out of Greymouth on the West Coast. The cards were stacked for Peter to end up working in the mine once he turned eighteen, the age that boys could start work underground. But instead, after his schooling had finished at fifteen, aside from a short stint as an offsider for a local delivery truck owner and then as a 'whistle-boy' in the logging industry, shearing was Peter's next job. The apple, perhaps, had fallen not too far from the tree – Peter recalls his father had done a bit of shearing for Tui Hart from Blenheim way back in his younger days.

Peter's childhood was filled with possum hunting, eeling, white-baiting, catching crawlies or 'yabbies', swimming in the Grey River, raiding the neighbourhood fruit trees, and running around playing cowboys and Indians and Davy Crockett with possum-skin hats and bows and arrows. School featured, but did not rate; the highlight, Peter jokes, was leaving. He attended the Marist School in Greymouth, where he enjoyed woodwork classes and sport but felt they 'definitely overdid it with religious studies'.

With preliminary background covered, Peter launches into early shearing memories at a great rate of knots and with much laughter:

'So the old man used to drink at the Dobson Hotel, and the publican at the time was Tom Heath, an old machine shearer from down here at Morven, Waimate. His son Jimmy was working with the Karaitianas and he was a blade shearer, and the old man did a bit of blade shearing, and . . . said to him, "Is there any chance of getting Peter a job over in Canterbury? Cos they're not taking any more boys on in the coalmine and he doesn't like the bush that much." . . . So I ended up getting a job with Jim Heath. He got me the job. He interviewed me in the Dobson Hotel – oh God, he put me through the third degree . . . I was about sixteen. He sat me down in the lounge one day, one Sunday morning or Saturday morning; whatever it was it was just him and me, and then – he's only twenty-one himself – he's telling me, "Yes, I can get you a job, but when you come over you've gotta keep your mouth shut and your eyes open, speak when you're spoken to, and when you're asked to do something you do it, and then you'll get to the stage where you won't have to be asked – you'll just use your head and go and do it." I said, "Go and do what?" He said, "Well, light the copper, get the hot water going, and if you've gotta bring some sheep up cos the pens are empty and it's smoko time, you hop in, tighten the sheep up, turn the grindstone, you know, at night, and all that sort of carry-on, and that's how you'll learn and – and – don't be smart, and have good manners."

'So that's basically what it was. And then I got off the train at Rolleston and he picked me up in his '36 Chev . . . The first pub we stopped at was Dunsandel; we went in there and he had a beer and I had a beer, and then the next pub we stopped at was Rakaia, and he had a beer and I had a beer, and then we ended up at Methven, oh God, and I got quite blimmin' drunk . . . and I went to bed that night and the room was spinning round. I think I spewed out the window – only sixteen!

'And then the next day we went down to his friend's place, a Māori bloke. And the sheep were wet, and he said, "I'm doing some washing. Hang my clothes out, would ya?" So I went out and I hung the clothes out . . . well, he come out and he went mental at me, he gave me a clip round the ear: "That's not how you peg clothes out on the clothesline! Didn't you ever watch your mother peg the clothes out on the clothesline? You don't use a peg for every tea towel or sock, you put 'em together so instead of using four pegs you only use three, and you go along the line. Use your head," he said. Little things like that, you know . . . it's just the journey of life: you pick up things. "Yes Jim, yes Jim." So I had to take all the clothes off and peg 'em out properly.

'So that was the first lesson; and then the next day the sheep were dry, so we ended up at Anama, somewhere just out of Methven, before we went up the Rakaia Gorge . . . There was Jimmy Coutts and George Karaitiana, and the sheep were a bit damp . . . so they went and played golf in the afternoon – they got a shedful [of sheep] that night, oh it was unreal . . . and the farmer said, "What, shearers playing golf, what's the world coming to?" And George was down to about seven handicap, he was a real good golfer, George Karaitiana. I was the caddy for the day.

'But the next day we finished the shed, and I think then we went up to Glenariffe up the Rakaia Gorge. I was a rouseabout; I was sweeping the board and picking the fleeces up – no girls, just all men, all boys – and the very first day Jim Heath said to me, "Come on now, come on now boy, have a go," and I shore the last leg of his sheep at smoko time or at lunchtime. And then as the days and the weeks went on, he'd leave a wee bit more until I could shear the last side virtually. And that's how you learned in them days: you shore the last leg and then you'd say to Jim, "Well, you go and grind your shears, I'll finish off your sheep for you," or, "You go and have a wash-up for your smoko," so the shearers used to like you doing the last leg . . .

'When you start your first season as a shed hand, by the end of the season you should be able to shear a sheep in five minutes. You go over to the shed at night and you practise cos you're way out in the country – you might be way out at Craigieburn or Mount White or somewhere, there's nothing else to do, the pub's too far away – and the men go to the shed and they're doing up pairs of shears and buggering around, you're turning the grindstone and then you've got an old pair of shears that you bugger around with, and if there's six shearers you'll shear six sheep at night, and you'll put the shorn sheep back in the pen, and you'll do the wool: you'll not leave the wool on the floor for the wool classer to do in the morning, so you roll the wool up and have it all skirted and nicely done . . .

'So that's how you practise, and while you're in the shed and you're practising at night the gun shearers will be there doing their shears up, and they've had a couple of beers after the meal at night and it's daylight saving and they can't get to sleep, so they'll teach you – bloody good buggers – they teach you how to shear, and everybody has a bit of an input, the whole team. Most of them are all pretty good, cos you shear a sheep for them so when they start in the morning they got a sheep out the porthole . . . Your goal is to be able to shear a sheep in five minutes, and that's the end of your first season as a rouseabout.'

When Peter asked for a stand, it was Paul Karaitiana who backed him.

'I went and seen George Karaitiana and his brother Paul was there, and I did my rouseabout with George; and I said, "I don't know, what about a learner's stand?" He said, "Righto, we'll give you a learner's stand and we'll teach you how to shear properly and then you'll bugger off and shear for someone else." I said, "I won't do that." He said, "We've taught other learners and as soon as they've got going they've gone with somebody else." And then Paul piped up and he said, "Don't worry about it, you can come with me, Peter." So he gave me the job . . . and I got a learner's pen and I did fifty for my first day, so that was like ten minutes a bloody sheep; bloody near killed me, I never worked so hard in all my life, and old Pop [Robert] Karaitiana was watching me . . . I had a face on me like a beetroot and me shears were blunt and oh God, I had big blisters on me hand, on me fingers. I carried on and I remember Ray Talbot took me shears off me cos he was there, him and Jimmy Coutts; he ground me shears for me and I went a lot better then, cos I'd throw me shears in and they'd come bouncing back out again they were that blunt; but anyway, we got through it, finished the shed.'

Peter recalls being at Ensors' Glenariffe station when he reached his first 100. It was Jimmy Coutts' words of warning that stuck in his mind:

'When I did my 101 Jimmy Coutts says, "Well you've done 101, tomorrow you'll shear 102," and I wasn't allowed to drop back and it bloody near killed me, it really did . . .'

The sheep are not shorn so closely when using blades, which means the grease (lanolin) is not lifted. The extra wool and grease give the sheep greater protection from the elements.

There was always a shout when someone got a tally, so Peter's cheque of eight pound ten duly got written and sent by the mail truck down to the Methven pub.

'There was no phone in them days. Mrs Ensor used to call through on the radio, and the cheque was sent down to Stewie at the Brown Pub who knew all the shearers, and he sent up eight pound ten's worth of beer, so that was on the weekend cos we always knocked off on Saturday morning, or some of them knocked off Friday night . . . and on the Saturday afternoon we used to do our washing and we'd start again on the Monday morning. But we had the party cos they all stayed on the station – they're all station stickers – and so we had the party Saturday night. So they drunk all the beer . . . but a funny thing that particular Saturday night . . . we got to about eleven o'clock at night and everybody's getting a wee bit hungry, so Bill Karaitiana said to me, "Hop down to the killing house son and grab half a side of mutton," so, "Oh yeah, righto," I went down and got the chopping block and the tomahawk beside the fire (cos it was an open fire in them days); he got the side of mutton and he cut it up with the tomahawk, and he got the fireguard and he put it on the coals of the fire and he threw the chops on and all the meat, and it was the best barbecue I've ever had in my life. We were eating there till about one o'clock in the morning – two o'clock – the half side of meat, and we had some salt, and it was beautiful. Yeah, that was my first barbecue I ever been to, and it was the open fire with the fire screen . . .'

When asked about the accommodation they stayed in Peter immediately remembers the beds, and what it was like with the quarters full of men:

'The beds had big hollows in them, and we used to put boards under the beds and that sort of carry-on. They were kapok mattresses, a bit lumpy and overall it was pretty good; I mean, there was probably too many men in a room at times, you know . . . you sit on the bed and have a yarn after your day's work. You have a beer and then you go and have your shower and you'd have your meal and you go to bed; you're tired, you sleep. You put a couple of nails in the wall, and you hang your breeches up.'

The mention of breeches leads Peter on to talk about how he remembers everyone dressing:

'And sometimes in the old days the shearers they used the same pair of woollen pants, and they had 'em on all week, old army pants . . . No, they wore tweed; that's the word I was looking for. They wore tweed pants, and we used to go to second-hand shops and get old suit pants; old wool before shearing jeans come out. You never used to shear in jeans: they were cold on you. But these new shearing pants these days, they're double-layered and everything, and they even got wool in some of them. Corduroys; we used to wear corduroys and tweed. And bowyangs? Yes, we'd see a lot of the old fellas cos it stops your pants pulling down on your hips. Even later on in life: in the last few years I've been wearing braces because me stomach's too big and I can't do the top button up. So you wear braces and your bowyangs and it keeps you all together . . .

'And we always used to use a jersey round our back to keep it warm. Especially in some of the portholes the wind would come in on your back, and your back's like a railway iron: it shrinks in the cold and heats up and stretches in the heat, so the idea is to keep your back warm all the time. There's nothing worse if you get a cold in the back, and you see young fellas in their pants with the crack of their arse showing and that's not good. I used to tell them off . . . Always a woollen singlet – never wore underpants, just a big long woollen singlet that went down round your bum cos the underpants with the elastic gets sweat in it and goes cold, . . . But in saying that, a lot of old shearers in the old days used to wear long johns under their shearing pants, especially if the sheep were a wee bit damp or it was really cold.'

Once Peter's 'apprenticeship' was over, he stuck with blade shearing as it suited his preferred work pattern, variety being of the essence: 'You get sick of looking at sheep's bums'. For over twenty seasons he was at the freezing works from early November through to the first week in March, when he would go opening oysters in Christchurch: 'Three hundred sacks a day come up from Bluff.' He won the world speed record for opening 100 oysters in 1974. After the oysters he would take up the blades again from around June or July. For a good number of those seasons he was a blade-shearing contractor.

Without a doubt, though, it was the competitive side of shearing that energised Peter; breaking the blade-shearing lamb record, which has stood for over forty years, is up there on his highlight reel. In 1975 he recalls getting a whiff of the potential for a record working with Les Smith:

'I did a big tally down at Mount Somers at Colin Greig's place, Flat Rock, and Les Smith and I shore there together. I did 274 – this is in a eight-hour day – and Les did 272, I think it was. He had me wrung out that day, but I knew I could step up . . .'

So the following year, fit and strong, he set himself up for the record attempt.

'It was at Colin Gallagher's place, it belongs to Blair now, a place called Rangiatea at the back of Mount Somers; that was in 1976 and I was only twenty-eight, but I did it in nine hours, and with all the judges and timekeepers it was official . . . I had a learner with me, Alan Archibald: he did 213 for the day and I did 353. I had two hundred out at lunchtime and I was doing forty an hour. These days a lot of shearers would have liked to beat my record and have a go, but unless they can step up to forty an hour they're wasting their time. Those sheep that I shore that day at Colin Gallagher's place, they were five-month-old Perendale–Romney cross lambs. It was a really hot day the day before, and the day I did it was drizzly, which suited me. But they never had a machine on them, and they weren't prepared; I went through the crutch and everything. They were full wool – people can't believe they were full wool – but I was just super-fit and had a mindset. Alan Norman was there that day, and Bert Loffhagen; they were the two judges. A JP was the timekeeper, and Mally Aldridge was the . . . manager . . . he got the local *Guardian* newspaper from Ashburton and they took a few photos, and then we all went down to the Mount Somers pub and had a few drinks that night.

Members of Mike Bool's shearing gang working at Mt Hay Station. With no downtube to restrict their movements a blade shearer can move around the sheep. Today it is a rarity to find blade shearers in the sheds of a high-country station.

'Yeah, I broke the world record. It's all in the top two inches, and you mustn't panic. I remember I gapped me shears, but I had a spare pair, and you were only allowed to have two, so that was quite good. Well, I had a small gap like I hit a stick or something or a little stone and then I went down behind me cockspur – I could have just spun out, but I kept me cool and carried on, and I remember I did forty-three the first hour before breakfast. I started at half past five, then we knocked off at half past six and I had forty-three out, so I was well on me way to 360. After smoko I did another eighty, and then . . . seventy-seven or something, so I did have the two hundred out by lunchtime. In the afternoon I knocked off every hour just to wipe me face and have a drink. I did have a shower at lunchtime and changed me clothes so I was fresh straight after lunch. It was the only time I've ever done that in me life . . . you'd think I'd been thrown in a creek, I was just – the sweat, it was unbelievable. The wife and the kids were all at home, and they were giving reports on the radio; they were listening to 3ZB in Christchurch and it was quite a buzz, really. I suppose records are made to be broken, but mine's been held now for quite a few years or over forty, and I'm very proud of that . . .'

On reflection, Peter thinks his Golden Shears win in 1975 eclipses the world record:

'I think winning the Golden Shears might have just clipped it. Your first big show, you know what I mean; there was something special about that . . . I won Waimate previous to that, but I think it was because George Karaitiana had won the Golden Shears, and he died in 1974 and he never even seen me win it, and he taught me – him and Paul. Never mind: it was a proud moment. I was very proud when I broke the world record, and then I won the world championships in Masterton in 1980, so I was right at the top of the game. Then they made me a Master Shearer, so I was very, very, very happy . . .'

Terrace Station, Craigieburn, Benmore, Mount White, Castle Hill, Erewhon and many more stations besides are mentioned as Peter gallops along, and the names of shearers he worked with swirl in and out of stories and accounts. He is generous with his praise of others and lists those he has been proud to compete against: Peter Burnett, Alan Norman, Sam Dobson, Sno Roffey, John Kennedy, Allen Gemmell, Noel Handley, Bill Michelle, Brian Thomson, Peter Race and Dinny Dobbs. And those he has done the hard yards with in the sheds he rates highly.

The shed shearers who were tough and didn't get beaten much were George Karaitiana, Jim Coutts, Alex Macdonald, who he rates as the fastest, Ray Harris, Bruce Davidson, Colin Cameron and a young one – Shane Casserly – who is 'best about at the moment'.

'And then there's Les Smith, Joe Love, Kerry Nolan, Ross Kelman, Peter Channings . . . the Scotts had a couple of teams going and they had a guy by the name of Dave Hamilton who was unbelievable: he made a beautiful job and just kept going day in and day out, and Graeme Rangi shore with him. And of course there's the evergreen Donny Hammond, Les Richards and a lot more like Mike Bool, Bryan Holden, Alan Butcher, Dave McMurray, Ronnie Hill, Baldy Bruce, Graeme Weaver, young Terry O'Donald and the Thomson brothers Ray and Brian. There were old ones, Arthur Sykes, Bert Loffhagen, Bill Johnson way back in the old days, and Jim Hura the big Māori shearer that used to hit first and ask questions later; he was a hard case. And, of course, you can't forget the great gear men Jimmy Deere and Phil Oldfield.'

So keen is Peter to acknowledge all those he worked with that he is very apologetic when he realises there are names he cannot bring to mind. A particularly good cook even rates a special mention:

'Back in the sixties and seventies when I was shearing for the Karaitianas and then . . . Donny Hammond, cooks were pretty good, but one really stands out – he was a great man. His name was Ross Wilson. He was single, but he liked a drop or two and he was good for no more than about a fortnight, three weeks, then he'd be looking for the bottle or he'd make up his own concoctions or whatever. He had to go away, you know; but anyway, while he was on the job he was unbelievable. He used to get around in his white apron and his hat, and he always had a beer on the table during the day. But he used to get up probably half past three in the morning and he'd prepare all the smokos, he'd prepare breakfast and lunch, and it was unbelievable – everything was spotless and prepared. He could cook anything. They say that he was a head chef on the *Queen Mary*. Yeah, he was unbelievable: he used to cook beautiful roasts, and the puddings were so fancy . . . plum duffs and everything. Oh, it was unbelievable; he was a great pastry cook, and you wouldn't know what you were getting in your smoko tray. But what happened with old Ross, he went to town and drowned in the Avon River, believe it or not. It was so sad. We were shearing up at Ben McLeod, with a couple of Māori fellas working

Suzy the feral sheep being shorn by Peter Casserly in Masterton. Peter had shorn Shrek the sheep whose fleece was far heavier than Suzy's. Suzy's fleece, however, was the longest length of wool, at 550mm, that Peter had ever shorn. *Wairarapa Times-Age,* 15 October 2018

in the team, and . . . cos he'd prepared all the meals so early in the morning, he'd go down to the creek and try to find watercress or pūhā, and he'd cook these boil-ups. He used to let the pot go cold and take all the fat off . . . and he used to put in the pumpkin and potatoes and throw the odd onion in; in went the watercress or pūhā. And, oh man, I can still remember how good they were – but the Māori boys wouldn't eat it. "Oh bugger that, no, he's taken all the fat out." So they'd eat the roasts and all the Pākehā were eating the boil-ups. It was quite funny: oh no, the Māori boys said the best part of the meat was the fat. I have the odd boil-up now and I still take the fat off. But no, Ross was a great cook.'

Peter has worn many hats over the years, and at one stage he owned the Omarama Hotel. It's easy to imagine him leaning on the bar, sharing a pint with the locals and regaling them all with his shearing stories. With great glee and hilarity, he launches into one:

'We were shearing away up at Glenfalloch and we used to go up there for the whole week – sometimes you were up there for more than a week. It's an isolated place, right at the end of the road up the Rakaia Gorge, and they were lovely people to shear for, the Todhunters. Across the river was Jim Morris at Manuka Point, and he was quite a famous person. He had size fourteen boots, a big tall fella, and he'd come across and see us now and again and have a yarn cos he was so isolated over there. Anyway, someone said, "Who's that guy?" I said, "It's Jim Morris." "Where does he live?" "Oh," I said, "he lives across at Manuka Point." "Oh yeah." And I said, "Yeah, he runs a picture theatre over there; they have pictures there every Wednesday night." "You're joking," they said. This was on a Monday, and then of course all day Tuesday we're saying, "We got the pictures on Wednesday night and that'll be good." And then Frank Lamming was there with us – he was a hard case – and he said to one of the rouseabouts, "Look, here's a couple of dollars: would you buy me a Crunchie Bar and a Buzz Bar and bring it back for me? Cos they got a bit of a tuck shop over there, too." And so these young rouseabouts – Terry O'Donald was one, he's a hard case, a shearer now, he was only young – they all got dressed up and asked, "What time?" "Oh, Tom Todhunter goes over and he'll be leaving about half past seven." So they all had their meals and put their best clothes on and away they went up to Tom and Prue's, and they knocked on the door. Tom come to the door, and they said, "We're ready." He said, "Ready for what?" "We're ready to go to the pictures." He said, "What pictures?" And they said, "Over at Manuka Point." He said, "I think those bloody shearers are having you on down there." So they were wild. Anyway, they come back, and of course we were laughing like hell; but I said to them, "Don't worry, don't worry, we'll have a few beers, and we've got the fire going, we'll have a barbecue, we've got some venison." So they settled down and we had a few beers. And anyway, out in the tree was a leg of horsemeat for dog tucker, so we shot out there and got this leg and we cut a couple of big lumps off it and cooked it up on the open fire. I didn't use a fire screen – we had an old pan ... They had salt and pepper and got some bread and butter from the cook shop, so they got stuck into this horsemeat. I didn't cut it too thickly, and they thought it was wonderful ... Just one of the stupid little things we used to do in them days.'

The only time Peter slows his pace is when asked why he stayed shearing. He took time to ponder the answer and explain:

'Well, I think it's like a muttonbird or an oystercatcher: you just keep coming back season after season. It's something that draws you back, like the muttonbirders and the oyster openers and the freezing workers ... it's the camaraderie of the friends you've met, and the good money, and probably that's all, you know. I suppose I could have become a truck driver or something like that; but what I loved about shearing was you only got paid for what you did, so if you were good at it you got well paid. On average, when it's all taken in – in the old days, you only averaged three days a week because of the rain and one thing and the other (now it's a bit different with the covered-in yards) – but you made more money in three days than your mates in town where they were working for five. When I started, on eight pound ten a hundred in 1965, my schoolmates were three pound a week as an apprentice mechanic; so I done a season shearing and I went home in a Mark 1 Zephyr, and me mates on the West Coast were still riding push-bikes. So, next thing, I had a girl in the car, the elbow out the window and a crew cut haircut – and I'm still married to that girl: Gloria and I have been married for fifty-two years and have three daughters and eight grandchildren.'

Peter had a long and illustrious shearing career, using the seasonal nature of the work to create a routine that offered variety. Someone else who has had a notable shearing career, but just not one in the spotlight, is Kelly Hokianga from Raupunga in Hawkes Bay. When someone tells you they've shorn nearly three million sheep you know you're talking with an experienced shearer, a fast shearer; but for Kelly being fast wasn't enough: he wanted to be a 'good' shearer.

'I'll tell you: in my opinion there's three classes of shearers, I've told a few people this. There are good shearers, and there are fast shearers, and then there are arseholes . . . It takes a while to become a good shearer, it's quite easy to become a fast shearer. But to be fast, and good to everyone and kind, considerate and all those good things – helpful and teaching and all that sort of stuff, well – it takes a while to become one of them because while you're a fast shearer that's all you want to be . . . A fast shearer, he's just a fast shearer; there's nothing else until he becomes a good shearer, and that takes a wee while. As for the arseholes: if you want to find out if you're an arsehole you just ask the shed hands, they'll tell you. They'll tell you whether you are or not. I'll just leave it at that.'

Kelly Hokianga –
Good shearers, fast
shearers and arseholes

With over fifty years of shearing under his belt, Kelly has earned the right to such philosophising. It is probably fair to say he has seen his share of each of the three classes throughout his decades in the industry. And a 'good' shearer is what Kelly has worked assiduously to be. He credits Johnny Winiata as the one who guided him when he started shearing professionally.

'Johnny Winiata, he put me on the right track – he kept me being tidy, taught me how to be nice to people. That's probably the best lesson I learned in those earlier days. Cos there's no use being quick and being mean to other people cos you get sorted out pretty well; you get put in the arsehole category and you don't get a job.'

At the time of the interview Kelly had recently turned seventy-one and had just finished his twenty-eighth season with Peter Lyon Shearing up at Mount Nicholas, where he still, according to Peter Lyon, tallied 200 merino ewes and 180 wethers a day. Kelly is taking a well-earned break, living out of his caravan and doing a bit of scuba diving. His love for diving has inspired him to stay fit, and it meant that he never pursued competitive shearing.

'I missed out on the competitions. All my friends went. Some of them became great shearers, like Eddy Reidy and Mac Potae . . . I got certified as a scuba diver in Napier and every spare moment I got I would be in the sea. I wouldn't go shearing, I wouldn't go to competitions; everyone else used to, but oh it was a different life down there.'

Over the years Kelly has been careful with his money, buying his first property at just eighteen, and though he is well set up for retirement and does not need to work, he is still drawn back to the boards. Lyon happily takes him on as ganger; his mana among young Māori lads is appreciated. Kelly says it's not quite so easy for him to shear these days (a little hard to believe, given his recent tallies), as his body is not quite what it used to be:

'But you never, never do as good as when you're younger because your movements, your body, it takes a lot of energy and strength. I've lost a lot of strength. And with all the hard work I did, I've had five hip replacements . . . two knees, I've worn my knees out . . . had my left elbow taken out – they've fused it.'

Kelly kept rebreaking his arm, and after three attempts to fix it a titanium rod was ordered from Germany. His orthopaedic surgeon, who he's now very friendly with, told him, 'They make Harley-Davidsons out of this stuff, so you won't break this one.' Kelly can no longer bend that arm, but, as he says, 'You make allowances. It's hard, but . . . you have to teach yourself new things. Of course it's not as good as the original one, but it's near enough.' Near enough, hopefully, to take Kelly to his milestone before he gives up shearing.

'I'll shear three million before I die. I was one of the first to shear two million . . . I was in England when I did that. It took me eighteen years to shear my first million, and it took me twenty-four years to shear my second million – I couldn't go any quicker because of all these broken limbs – so it'll take me till I die, and I'll just have to cruise along. Just keep going.'

Kelly's experiences could fill pages, especially his travel stories. He's travelled extensively, shearing in the United Kingdom, Europe, America and Australia. Robin Hood country – Sherwood Forest in England – is his favourite spot other than New Zealand. In the early days he honed his travel techniques to ensure ease of passage, entering under the radar to work in Britain by always travelling dressed 'in a shirt and tie, and I didn't look like a shearer – I looked more like an accountant. You had to bluff your way through customs. I did it for years and years.' It wasn't unusual for him to get upgraded to business class when he flew.

He laughs with the memory of his naive efforts to bring his hard-earned dollars back to New Zealand:

'The worst thing I ever did, the first time I went over there, I brought my money back in cash – in pounds. I had £17,000 in my man-bag . . . it was stuffed in there and inside my jacket and down through my track-pants. I came back through Heathrow, and they didn't scan you for that sort of thing in those days; I got away with it the first time, but I was backed up against the wall. Everyone looks like a robber . . . you're suspicious of everyone. Then I got to Auckland, and New Zealand's hard to get into especially with money; they've got the sign flashing, if you've got over $10,000 you have to declare it. I looked at all the custom officers and I saw this old Māori lady, she was one of them. I just ran up to her – I pretended I was limping – and she said, "What's the matter?" "I fell over in Scotland and twisted my ankle and it's quite sore." "Oh, just go through, you'll be right." So she let me through – she didn't even search my luggage. I was so lucky.'

Kelly's early Australian experiences coincided with the days of the Wide Comb dispute when his mate Aussie Carrington got shot.[*]

---

[*] The early 1980s saw the escalation of a long-simmering debate in Australia over the use by some shearers of a wider comb than the Australian Workers' Union allowed. Since New Zealanders used the wider comb as a matter of course and were working in Australia in greater numbers, they were perceived as undermining the Australian shearers' working conditions, creating animosity and even violence.

'It was horrific over there in the early seventies when we were getting shot. You'd go to the pub and it was just a big brawl, Kiwis against Aussies . . . I shore with a shotgun on my stand when I was there. I wasn't going to die for anyone. I would have shot someone if I saw someone walk in the door with a gun. Well, I think I would have – but thank goodness I didn't shoot anyone.'

Australian conditions were harsh: there could be 1400 kilometres between sheds, and the stations were huge – a paddock could be 1000 acres. With 50–60,000 sheep at a station the shearers could be at one shed for weeks.

'For every young Kiwi it's hard over there: the conditions are worse than anywhere else that I know. Some of the places there's no toilet – you've got a spade with toilet paper wrapped around the handle and that, and you just go out and dig your own hole. Honestly, that's what it was like in those early days.'

There were many trips to America as well. Teaching through the shearing schools opened doors for him.

'I used to help them out and teach some of the younger guys, so I went in there big time. Alan Barker . . . he was the leading shearing instructor in those days. He'd just write a bit of paper out for me, send it to his friend over in the US, my name and everything, and I got visas straight away. If you're well known in the shearing industry the opportunity will always be with you. I had opportunities to go anywhere I wanted. You had to behave, though. Your attitude's a big thing in shearing – a big thing in anything, really, but especially in shearing. If you've got a bad attitude you might as well stay home cos you won't get on.'

The shearing season in the States was brief, though; Kelly would shear for six weeks in Utah.

'You shear in the trailer, like a big passage. The sheep run up the races and then you just push the lever, there's no dragging; you push the lever with your foot and the sheep will drop down right where you are. If you're claustrophobic you'd be no good there. They still do that today.'

Temperatures were 30 degrees below zero, and even with the bearskin coat his Sioux friend Jake made for him Kelly was still cold. He laughs as he recalls sleeping out in Jake's tent, checking for polecats and not being so keen on Jake's cooking:

'I ate vegetables; I didn't trust him with his meat. Cos he'd cook a skunk, but he'd have so many herbs in it – if you had skunk on its own you wouldn't eat it cos it is shocking! I smelt it and said, "I'm not eating that – what the hell's that?" "It's skunk." "I'm not eating that." I just refused! . . . But I do eat a bit of bear. When you see a bear skinned, hanging under a tree, it looks like a muscly oversized human. He'd just cut bits off and leave it hanging there for days. Flies used to get around – just scrape the maggots off. Ugh. He was healthy as – never went to the doctor.'

Despite all his travel adventures Kelly remembers how he never wanted to be a shearer. Born in Wairoa in 1946, he had wanted to be a carpenter; but with parents who met in the woolshed and a grandfather, Teoti Hokianga, who had once been a big Hawkes Bay contractor, it was little wonder shearing became his life. There was nothing sugar-coated about Kelly's upbringing. He comments that in his day it was normal to have families of twelve or more. He jokes, 'We were a small family – we only had eleven.' But he also comments that, within his immediate family, 'it wasn't the best environment. If you've seen *Once Were Warriors*, we were all brought up around that sort of thing; that wasn't new to us when we saw that movie, it was just natural life for us.'

Early life was spent living between a sawmilling community at Tūtira in the backblocks of Hawkes Bay where his father worked, and grandfather Teoti's Raupunga farm. As was common in isolated environments at the time, amenities were at best basic. At the house in Tūtira, Kelly remembers, 'we had to go down to the creek to get our water and bathe in an aluminium tub. Of course, there was no sewerage; it was all long-drops. And most of our long-drops had two seats, like you could go in pairs.' Kelly stops and laughs, adding, 'I think it was Dad's idea to save time, I s'pose.' His grandparents' place he remembers as a big seven-bedroom house. 'It wasn't a flash house, but it was a huge house.' There was no power and the 'flashest' item was the crank-handle phone on the old party line.

Kelly was expected to help on the farm at Raupunga and remembers having to go out with the men in the school holidays to cut scrub. 'You'd tie a hunk of cold meat on your belt. You'd just keep chopping all day and just have a bit of a bite and carry on.' He laughs again. 'We thought it was normal.'

Aged around eight, Kelly was made to take up the handpiece. With his grandfather stocking 2500 breeding ewes plus hoggets and wethers on his property, it was a big ask for the young boys in the family.

'I didn't want to hold one of them. We used to have a three-stand woolshed quite close to our house, and when shearing or crutching time came three guys used to come up; sometimes Dad got on and helped but he'd make us do the dagging and all that. They'd just set the handpiece up for us and show us a few things, and you had to go and do it otherwise you'd get in trouble.'

Despite doing well at school, Kelly left at the age of fourteen. When his grandfather died, a man he idolised, he refused to continue his schooling and his working life began. He moved to Palmerston North, where he got a job shearing with his father and a cousin – but only, he quips, because he wanted to be a 'boss'.

'They were both shearing, and in the first year I didn't shear – I was just helping with the wool rousieing and that – and I thought, "I hate this." I hated it, so I said, "I might learn how to shear," cos all the shearers were bosses – like they'd boss everyone around and got all the good stuff. I said, "I want to be one of those." I just wanted to be one of the bosses; so that's how I became a shearer at a very young age. The shearers never had to do any of the lowly jobs: they were allowed to do anything they wanted, and I liked that. In those days, I couldn't even ask the shearers anything – you'd get a whack over the ears. Like you'd sort of have to be a mind-reader. So I never used to say anything: I'd just do what I thought was right and then run for it. I actually learned to swear in the woolshed; I didn't learn anywhere else because it was shocking.'

Those in the shed with him on his first day were Johnny Winiata, Lofty Crawford, and brothers Jim and Tex Wilson. He remembers without joy that he was made to run around after them all:

'By the end of a month I almost wished I didn't go. A typical day was just shed-handing . . . most of it's just filling the pens, like the sheepo . . . they'd yell out "sheepo" so you'd go and fill that up and go and help the presser. And then after the first month I said I better start from the bottom, so I asked for a presser's job, not a shed-handing job. So they let me do that. I used to work very hard on the press, right up to about six o'clock at night while everyone else was drinking beer and doing their gear up. And one guy said to me, "You should learn to shear properly." And I said, "Oh yeah? I will." He said, "Come here, I'll show you." So, he showed me the first sheep I ever shore properly. And I quite liked it. He said, "You want to come over here at nights and just take a few bellies off and get used to the feel of the handpiece." I said, "I know how to take the belly off." I was lost after that. But I kept doing that every night and he kept coming over and showing me different parts of the sheep and that's where it started for me. I kept going there every night and I was getting better and better and better. I never forget one time, though. In those days the lights were quite dim and we were shearing in a place called Harry Barber's, just out of Palmerston, and I took two hundred bellies off; I just took the bellies off because I wanted to take them off as fast I could, so I cut forty tits off. And I got a bashing from Dad after that night. I wasn't allowed to go to the woolshed any more after that night, but I learned not to cut tits off pretty quick. Which was fair enough . . . you cut a tit off one of those, they're worth nothing.'

His first 100 came quickly, in about three days.

'I was so determined to do that and get on with it. I just didn't want to go back shed-handing, so I learned a lot of things. I didn't listen very well, but some of the older guys that were trying to teach me, I knew they weren't all that good either – like, two months later, they were doing way less than I was doing, so I was glad I didn't listen.'

Kelly got his first stand at fifteen.

'Oh, it was incredibly hard, especially how I shore at that early stage: it was all brawn and no brains, I should say. I wouldn't listen to anyone – I felt I couldn't listen to anyone because to do what they were trying to teach me I had to be more skilful and more repetitive and I couldn't do that unless I was working hard, so it worked for me. Cos a few months after that I got the idea of what I needed to do without anyone telling me . . . I got pretty quick in those days, very quick.

'I did my first five hundred when I was fifteen. I was pleased. There were quite a few of us from back in Raupunga that did that; I'm not the only one. There was a thirteen-year-old last year that did four hundred from there. Oh no, you learn how to work back there; there's not much skill in it, it's just all raw guts. It tells in the end, especially when you get older; you learn it's not easy, and you're prepared. Some people can't hack it: they just give up and go get a job in the freezing works or the roadworks, something a bit easier.

'I done my first six hundred when I was eighteen. I had to wait till I was twenty-one before I did seven hundred . . . that's three years. I was angry with myself because I never quite made it. It was only because I wasn't fit enough, but I didn't recognise it. No one was into fitness in those days. Hardest thing you ever did was walk over to the cookhouse for tea. You had no gyms in our day . . .'

Kelly comments on how the importance of a good diet is recognised today, too:

'Diet is the big thing in shearing these days, and you get the best food available. In our day our main diet was boiled mutton and spuds. And for a change it was spuds and boiled mutton. You never had much choice . . . my dad, he hated roasts, so we had to do what he wanted. One day, the cook put a roast mutton into the oven – even my brother was angry with him that day, cos he pulled it out half-cooked and threw it in the pot and filled it up with water and boiled it. I was so angry, but I didn't dare say anything. We lived on boiled food for most of my younger shearing days.'

To keep improving Kelly knew he had to be up with the best. 'When you're a shearer – even today they do it – if you want to get better you have to go shear with better shearers, otherwise they'll waste your time. You'll never get better unless you're competitive with someone better than yourself.' So he started working for Colin Kendal, then Keith Sandbrook, who was a bigger contractor working around Masterton. Then when Alan Parker started a run in Palmerston North, he moved to him. Kelly thinks back to those early days: 'I battled a lot of good shearers . . . Joe Philips, Luke Campbell – he was left-handed, probably the best in the world at the time.'
    Kelly laughs at the memory of his first handpiece:

'It was a really thin, old-looking thing; I don't know how I ever shore with it. Mine was a Lister – narrow gear, of course . . . and you'd have a bucket of water beside your stand cos sometimes it would get that hot you'd just throw it in there to cool it down. A lot of people did that in the old days. If you told younger guys today, they'd laugh at you, "Oh rubbish." You couldn't hold on to it, it was that hot – they weren't very well made, the steel was good, and . . . because it was good steel it wouldn't cool down that well; so the bucket of water – you'd put it in there, then dry it off and put more oil around it and carry on. Your hands were blistered.'

When Kelly started, the shearing season was shorter, so he didn't need quite so much gear. 'I would only have five combs and maybe twenty cutters . . . We had to reshape the teeth on all our combs; in the earlier days you had to be a sort of scientist, to develop your own shapes and the ends of the combs.' Some shearers used the nail hole in a horseshoe to reshape the last tooth of the comb. 'I cut off my dad's old .22 [rifle] he had and drilled a hole in it – I just put the top of the .22 over the comb. I'd heat it up first, of course, and just pull it over just like that . . . and that was better than trying to hold a horseshoe.' He remembers how all his gear could fit into one of his moccasins.

Shearing culture has always had a reputation for being pretty rough, and it's fair to say in Kelly's earlier days life in the gangs could be harsh. On stay-outs in the gangs where alcohol was prevalent, ugly scenes could eventuate and frayed relationships were challenged. As Kelly says, 'People bring their worries to work.' Witnessing the sorting out of 'worries' was another reason why Kelly wanted to be a 'good' shearer.

'The culture has changed. In the North Island, the shearers had most of the say. If you were a rousie or a presser, you just fell in line behind the shearer. That was the biggest culture – that and food, of course. It's just, "I'm a shearer, you're a rousie or a presser, you go last." But nowadays you just fall in – you say, "Oh well, I'll just wait till you go." Some of the girls jump in before us; if you did that in our younger days you'd get a kick up the backside. You wouldn't be well liked. But it has changed: shearers are miles better this day and age, they've got more consideration . . . today they ask the younger ones, the pressers or the learners, if they *don't mind* doing this. They don't say, "Hey, come over here." They're quite civilised these days.'

The determination to be a good shearer has served Kelly well.

'In the shearing industry you get well known – you only have to pick up the phone when you come good . . . Contractors knew who I was and were quite happy to have me. They're looking for good staff – reliable, don't muck around, able to carry on working – that's all they want.'

Success has opened doors for Kelly, and he has relished the opportunities it has given him: the travel, the mates and the financial freedom to pursue pastimes outside of work. While Kelly is coming to the end of his shearing career, Sarah Higgins is another 'good' keen shearer just beginning hers. She is one of a growing number of women now shearing in the sheds.

Sarah Higgins – Catching the bug

'Oh, the bug had caught hold way back when I was fifteen and I was tryin' to shake it off I suppose. I had said to Chris Jones, "I wanna learn to shear, I wanna be able to do it." And he said, "I'm not teaching you until you're eighteen because your body's not finished growing yet and it's a bit hard on your body, so until you're older, don't do it." But the shearing shed environment had definitely caught on by then.'

At twenty-six years of age Sarah Higgins has packed more into her life already than some twice her age. After finishing a BComAg from Lincoln University and a Tectra certificate in wool technology, she worked with H Dawson Wool long enough to realise that working in an office didn't suit her. She travelled to the United Kingdom, where she woolhandled and shore; she's woolhandled in Germany, worked in South Australia and taken on shearing contracting in Marlborough in 2015. Awards are stacking up: she was the first person to win a woolhandling title and shearing title at the Golden Shears, winning a novice shearing title and a junior woolhandling title; she still competes in Senior shearing and Open class woolhandling competitions, winning the prize for Top Quality Shearer in the Senior grade heats at the Golden Shears in 2018, as well as Best Quality Woman Shearer. Recently, too, she became a member of the New Zealand Shearing Contractors Association executive committee.

Sarah's energy and enthusiasm for the shearing and wool industries are grounded in her early childhood, spent growing up on her grandparents' sheep and beef farm at Havelock. The woolshed at shearing time held great appeal for Sarah, although to the shearers she may well have been a nuisance.

'Since I've started shearing, one of the guys I bought the run from and who shears for me now was shearing there when I was little, and he's told me I was a pest. I used to sit on the stand behind him and just never stop talking all day in the shearing shed. I stopped a whole shed working one day when they thought I'd gone down to the river, but I'd actually gone up to the house to see Nana do the cooking.'

Sarah's mother Fiona, who had been a wool classer, introduced her to woolhandling at fourteen. Picking up weekend and holiday work in the sheds, Sarah was soon trying to get days off school.

'They start at seven-thirty and it's only one on one there – one shearer to one woolhandler – which makes it a bit easier; but just sorting all the wool and keeping up with the shearers is the biggest thing. Talking flat out as well, telling stories – I was always quite good at talking, makes the time go faster – and that's probably what I enjoyed about the work as much as anything else . . . You'd talk about what's going on at school, and the common questions are always, "Have you got a boyfriend yet?" or "What are you going to do when you leave school?" "When's your holidays?" – just common stuff, really.'

Sarah reflects on what it was like to learn to throw a fleece:

'When you're learning, it definitely takes a while to get your head around it, and there are so many little things you can do which make a big difference. It takes years of doing it, and talking to different people who have all got slightly different techniques, to work out what works best for you. Probably doing the competitions helped a lot in mastering it, because ultimately you want the fleece to land as flat as possible – no folds and not off the side of the table. It increases the efficiency of the shed a lot if you can get it right, but it takes years; now we're contracting and teaching new woolhandlers you sort of see it from the other side . . . but it does take a lot of time and practice.

'I probably learned a lot from just watching – especially the Open woolhandlers at competitions – and asking lots of questions. I particularly remember struggling with the neck, which would end up tucked under; and so it was about sort of flicking it more, or the way you set it up when you picked it up so that it would land flat. And I remember another day we were doing big merino rams and I had another woolhandler with me, and she couldn't lift them off the ground. She's like, "These are too heavy; I literally can't get this off the ground and throw it nicely on the table." It would have been getting up to 10 to 12 kilos, and when you're on a flat board and it's a big fleece, holding it properly to get it to throw well can be a bit difficult; but being a bit taller I found it easier – that and having longer arms.'

Sarah headed down to Canterbury after finishing at Marlborough Girls' College, recognising she could follow her love for agriculture and focus on the marketing side of wool at Lincoln University. It was at varsity that life in the sheds really wooed her. A mate, Ethan Pankhurst, who happened to be a shearer, saw her woolhandling ribbons on her dorm-room wall, and organised woolhandling work for her with contractor Barry Pullin.

'And that's where I started learning to shear. It's not uncommon for pressers and woolhandlers to say, "Oh, can I have a go," and they'll be like, "Yeah, do the last side," and so you learn backwards: you master the last side of the sheep, and then you work back towards doing the last half. I think I shore my first ever full sheep with some of the guys down there in Canterbury; and then in my second year of university I did my first learner shearing course, back up here in Blenheim. I skipped a whole week of university to come here and do the learner shearing course.'

What stayed with Sarah from her time working for Barry is the memory of some of the places she went. 'I saw places in Canterbury I never even knew existed, and . . . I got to go to so many places I had never been before. I went over to Banks Peninsula a couple of times, down to Little River and sheds down there, and up to Lake Coleridge.' Sarah liked it up at the lake, as it was 'just that far away – you're right up in the high country and you're shearing merinos up there, which I hadn't done a lot of; and there were times when we were shearing, and it was snowing outside . . .'

After finishing her degree and deciding against the office environment, Sarah took up opportunities to woolhandle and shear in England, Australia, Southland and, perhaps most challenging of all, Germany, where she was based near the Dutch border.

'We drove round in this old orange Volkswagen van that had five seats in it, and the back of it was all kitted out so the shearing boards were welded with plywood on top and folded in half. They were about a metre and half square, and that is what you shore on. It had a pole you pulled out and hung the machine on, and we had two of those and a generator. There were four of us: Simon and Rob shearing, I was woolhandling, and we had a guy who was catching. The farmer would set up portable yards, and you'd often have a big pen and your day shearing would be really slow. Often you just let the shorn sheep go in the same pen as the woolly ones, so you had to dig through – that was what the catcher was for: to catch the woolly sheep and pull them out for the shearer, because otherwise you wouldn't get any sheep shorn, chasing them round this big pen all the time. It's even more backward than England. Some days we managed to set up the pen with a gate: he would catch and drag them out, and then the shorn ones would be let go in the paddocks, so they didn't get boxed up. We struck a lot of farmers who didn't have the stock sense and didn't know how to set up yards properly to get them to run. They were never ready when we got to them, so we had to spend an hour helping them set up the yards and get the sheep in, and they never seemed to understand that we were under time pressure: even though their sheep were finished and they were done for the year, we had another day the next day, and the next . . . On a better day the boys might get close to doing two hundred each, but often we couldn't start early cos they weren't covered; it was outside. And this was late spring heading into summer, so often they were in lamb and some of them were lambing on the board. To be honest I would not go shearing in Germany myself; it was too hard. The sheep were full. They're massive, they're fat and they've got lambs in them, and they're sticky and it's not easy . . . often we'd have three mobs to do in a day and we'd always be packing up and moving.'

Sarah laughs at some of the ridiculous situations they encountered:

'It's all council-owned parks and public land where sheep are grazing, and one day we were shearing under a four-lane motorway bridge cos it was where we could get the sheep and they'd be dry, so we could start early in the morning. So we had this massive pen set up, and all day we were listening to the *kadunk*, *kadunk* of cars going over the bridge and there's people walking past, there's dogs everywhere and we're just there shearing sheep. You'd get people coming over who can't speak English and they're looking at what you're doing – I couldn't say much to them cos I couldn't speak any German.

'And the Green Party in their government wanted to save the wolves and had this big "bring back the wolf" campaign: you're not allowed to kill wolves, and the government will pay you for any sheep that a wolf has mauled, but they won't pay you for the ten that drowned in the creek cos the wolf spooked them. That's not considered "being killed by a wolf". So if you ever got rid of a wolf, you followed a three-S rule: shoot, shut up and shovel.'

When the opportunity to take up the runs of Marlborough shearers Chris Jones and Tony Nott came up in 2015, Sarah decided to grab it.

'I guess I took on a lot at the time because I wasn't just learning how to run, starting a business: I was learning to shear myself as well. And a lot of people did say, "Are you doing the right thing, Sarah? If you really want to be a shearer, you need to just be a shearer, and be able to travel and shear in other places . . . Don't get tied down here." But . . . for me the best opportunity was to start here and get my full-time stand. The only way to carry that on was if I took over myself, cos these two guys didn't want to carry on. I guess in hindsight I could have said to them, "Oh, I don't want to do the contract, I just want to shear, and I'm going to go elsewhere and travel"; but I took it on. I wanted to do the business side of things as well . . . and a lot of people probably looked at me and thought I was quite young, and I don't know, like, "What does she know about shearing?" But I'd done a lot through university and travelling, and I'd been around and worked for a few different contractors in a few different places.'

Sarah's run is diverse, with high country up the Awatere; it goes through Wairau and out to the Malborough Sounds, with shearing on Arapawa Island and way out at Anakoha. There are some merinos in her run, but mostly it's cross-bred second shear. With it just being a single gang, they're all a tight-knit group, with Sarah, her brother Duncan, Chris Jones and another mate shearing, plus a sister-in-law woolhandling and extras pulled in as needed. It's early days for Sarah yet. She is feeling her way and not ready to overcommit herself, although in the future she hopes to grow the business.

While female shearers are appearing more frequently around the sheds in New Zealand, and there are other female shearing contractors, they're not exactly a dime a dozen. And although Sarah hasn't encountered any resistance, there is occasional surprise at just how capable she is.

'We were shearing rams one day at a shed here and a shepherd didn't put any in my pen. I was like, "What are you doing?" and he's like, "Are you going to shear these?" and I'm like, "Yeah, well I'm not going to stand here and watch the boys do it." And once they see you shear and realise that you're quite capable, there's no discrimination whatsoever.'

She believes it's as much about the mental attitude as it is about strength.

'There's plenty of guys out there who get really tired and get really angry and grumpy and throw the towel in and walk out of the shed. It's a mental thing, it's not a physical thing. When you're tired and physically challenged it comes down to your mental ability, not how physically able you are. That's what it is. So there's no reason why a female should be any less mentally able than a male.'

Sarah admits to being pretty buggered by the end of the day, but ironically it's that very physicality that she loves about the job – 'the challenge and the constant learning'. Reflecting on what it is about shearing that keeps her going she replies, 'Everyone talks about the shearing bug, and you either have it or you don't. You either love working in the woolshed or you hate it . . . they talk about you catching the bug and then you can't shake it off.'

While Sarah Higgins found shearing irresistible from an early age, it's fair to say that Jack Luttrell, shepherd at the Matthews family's Waiorongomai Station on the shores of Lake Wairarapa, had no desperate desire to shear. But wartime manpower shortages meant that staff were called on to shear their flock; and with the incentive of being paid extra for shearing, he took on the challenge . . . with mixed results.

LEFT Not all countries have farming practices as well honed as New Zealand so shearers, when working overseas, need to be adaptable. Shearing under a four-lane motorway bridge in Germany is one of the more unusual settings for a day's shearing. *Sarah Higgins collection*

BELOW Before the leather or felt moccasin came on the market, shearers would make them out of the tops of the jute woolpacks – much to the farmers' consternation.

A woolshed is an expensive
asset that sits unused much of
the year. Waiorongomai.

Jack Luttrell – Filling the breach

'During the war years when shearers weren't so available they had two working there who'd been shearers: a chap named George Brickell and another chap named Muir. So, anybody interested from the shepherding staff was encouraged to be a shearer – and that was me because we got paid so it was an extra on top of our wage.'

The 'there' was Waiorongomai, a station on the western side of Lake Wairarapa, which has been in the Matthews family for seven generations. When a sixteen-year-old Jack Luttrell started working there, the 'supreme boss' was Mr RW Matthews, and the 7000-acre property had 3000 grazable acres stocking 1500 breeding stud ewes, as well as about a thousand ewe and ram hoggets. There was also a small Hereford stud. Jack had always wanted to work at the station, but his primary interest lay in the potential for hunting opportunities. 'Waiorongomai was always a dream place for me, with deer and pig, bush and lake.'

Talk of hunting aside, Jack laughs as he remembers himself as a young shepherd learning to shear.

'I had a terrible struggle learning how to hold them. There were two or three of us that were shepherds at Waiorongomai that they brought in – they didn't let us loose on any of the stud sheep; we had the killers to practise on, the dog tuckers . . . and if you lopped something off it didn't matter. You just learned by trial and error, and the good shearers – you watched them and picked up as many tricks as you could.' When asked what his first shorn sheep looked like, again Jack laughs: 'Oh, terrible! Bits cut out of it, bits hanging off round its eyes and ears, and underneath its legs looked more like a scarecrow . . .'

Jack recalls struggling to control the old Wolseley handpieces:

'The handpieces that were available, particularly the station ones, were old Wolseley handpieces that ran hot – terrible things to work with . . . the Coopers came along a bit later, they were better. The overhead gear had a big shaft right through and was driven off that, and had the up-and-down rope.'

He diverts to telling a story about shearing down at Doug Eglinton's property when they were short of shearers:

'I used to shear out on the coast and they had an old machine there, a Lister Blackstone engine, and it had double ropes: you pulled one to switch it on and one to switch it off. When they started up the engine in the morning it would take a while for it to get hot; it was an old petrol motor. So as soon as it started going, there were three chaps I was shearing with at the time and they would get a sheep out. If we started shearing before the machine got hotted up to drive the thing properly, the head shepherd chap would rush along the board and switch all our handpieces off. And you'd hear the old engine catching up and building up the revs, and as soon as you thought it was going loud enough you'd switch it back on again and it would slowly go. The chap used to run back and forth along the board; it was quite funny.'

Returning to his Waiorongomai memories, Jack recalls the full-fleece hoggets:

LEFT Waiorongomai woolshed, with extra bracing to help withstand the northwest gales. The wind is strong enough to move the steel beams inside the shed two to three inches.

BELOW Staff of Waiorongomai in the late 1950s enjoy a well-earned smoko break. *Waiorongomai collection*

'The Waiorongomai hogget fleeces, they had staples about eight to ten inches long . . . everybody hated shearing those full-wool hoggets . . . and when you opened them up their fleeces were so big and heavy they'd pull the skin out. Oh, terrible to shear them, they were shockers. You had to try and hold the fleece up so you didn't cut the skin off them all the time, and go slow. They were dreadful shearing, those hoggets.'

When asked how he felt after his first full day of shearing, Jack laughs yet again:

'Terrible! I'd completely had it, cos my back used to ache when I was learning to shear; it gave me hell. I think it was only the thought of the extra money I was earning that kept me going.'

Shearing only on Waiorongomai and at a couple of other local sheds, his first 100 took a while – 'probably two or three years, I think. Quite often when I was just working with sheep in the paddock, crutching or dagging them with a pair of shears, I'd be practising where I was going to put the handpiece on their body and practising holding them.' He was keen to improve his tallies, and while others shore 'quietly along' Jack always tried to shear as many as he could.

Not until internationally acclaimed Godfrey Bowen visited and carried out a public demonstration, having just made his world record – 456 sheep in nine hours on 6 January 1953 – did Jack appreciate the true art of shearing.

'Course, I was mesmerised by how he got the wool off . . . and for us self-taught people, you couldn't believe what you were watching. The way he did it. He really revolutionised shearing all over New Zealand, it set the standard . . . he made every blow count. When you went from the brisket to open 'em up, down, he made it look so easy . . . And he taught everybody to fill the comb – that was the thing, and to do away with second cuts and things. But he had a special method of holding the sheep . . . so that he virtually was moving all the time. If you watch these top shearers going now, the sheep's hardly in one place for a second. He showed people how to take the bellies off quicker and do the crutching and the legs, everything . . . the way he held the sheep made the long blow easier to do – that was my interpretation of it.'

Jack had only about five or six years' shearing when he became head shepherd at Waiorongomai. Though his shearing days were over, 'the staff were still doing it; it was a bit of a perk and it suited the way they ran their sheep: they didn't have the pressure of other shearers hanging around all the time and they didn't have to feed them or anything'. Jack left Waiorongomai in 1965 to go farming, but he thinks staff continued to shear on the property until the 1970s when Raymond Matthews was the owner. Nowadays, Raymond's son Charlie runs the station and John Hodder is the shearing contractor.

Just as Jack had taken up the challenge of shearing in a time of need, so, too, did Murray Grice, who was born and raised in Ashburton. A shift to his stepfather's farm in Hinds introduced him to farming and the shearing shed, but the handpiece was not seriously taken up until his brother, working on Terrace Station at Hororata, was short of shearers in the 1950s and roped his brother into filling a stand.

Terrace Station, once owned by New Zealand Premier Sir John Hall, is still home to his descendants. Built in 1869, the twenty-stand woolshed has seen thousands of sheep cross the board during its years of service. Today it stands quietly awaiting further restoration by the Terrace Station Charitable Trust.

Murray Grice –
An extra pair of hands

It is over six decades since eighty-five-year-old Murray Grice first saw the Terrace Station shed, but he retains a sense of pride in being part of its history. Sitting at the dining-room table at his home in Hororata he explains that a shortage of shearers meant his brother Reg, a shepherd at the station, was in desperate need of an extra pair of hands. Murray is not quite sure why there was a shortage, but he knows he filled the gap sometime in the 1950s.

'I'd gone and worked in other shearing sheds, but I'd never picked up a handpiece in another shearing shed, and me brother arrived down home one day and asked my stepfather, how busy were we? He probably said, "Not too bad," and Reg said, "Can Murray come up and give us a hand to shear? Cos we can't get shearers – we've just got one old Māori fella." I think his name was Tom. I can't remember the time of the year – I know it was before Christmas – and up I come. I come up on the Sunday cos we were shearing on the Monday. Reg said, "Come up and I'll show you the sheep and the shearing shed." So, I went down, and I'd never been in a shed so big – twenty portholes – and the shed is there today, was built in 1869, a massive shed. And a half past seven start . . . of course, I'm green and I've got to feel me way, and me brother had been a shearer and he was the shepherd there, and he kept showing me a wee bit more. I think I did sixty-eight the first day – I can't remember. I don't know how long it was before I got me hundred, but once I got me hundred it seemed to come easy after that.

'I think they were Corriedales – I'm pretty near sure they were, but I could be wrong – and every now and then there would be some wool thrown at me. I didn't know what it was, and I said, "Somebody's throwing wool at me, Reg." He said, "Are you pulling the bellies off, Murray?" I said, "No – why?" He said, "The wool classer's trying to tell you something: pull your bellies off." Cos I was just cutting the belly off and flipping it round, you see – leaving it on the fleece. You're supposed to pull it off. He was pulling it off and throwing it back at me . . . but he should have come and told me because I didn't know.'

Murray laughs at the memory, and chuckles some more as he talks of the wily shearer selecting the easiest sheep to shear:

'Me brother, as I say, was the shepherd, and he'd pen up; we were both catching up out of the one pen – this Tom, that old Māori fella, and I – and Reg said, "Get some of those bare-headed ones, Murray, he's taking all the good sheep on you." Course, I didn't know, I just went in and grabbed a sheep and took it out and shore it.'

As an aside, Murray recalls old Tom:

'He was pretty quiet, he was a real gentleman though; he died in that woolshed a few years later . . . died in the wool. Cos he used to come back and do odd jobs for them and a bit of shearing. And he died in the shed, he'd be quite happy doing that. I don't know how many years he shore there, but I never struck him again after the first year I come up to shear there.'

While he was a little light on shearing etiquette, Murray had at least shorn a few sheep. His stepfather, Lou Reverly, had a 1450-acre farm just out of Hinds where Murray lived with his mother and siblings. The farm stocked with 3000 ewes and 1000 replacement hoggets meant there was opportunity at shearing time to try out the handpiece. Like so many budding shearers before and after him, Murray first held a handpiece to crutch, and then progressed to trying to shear a sheep at smoko breaks. Mr Bennett from Hinds and George Parish from Geraldine were the shearers who travelled in each day.

'I used to try and shear one while they're having their smoko in that half-hour break; sometimes I'd get it done, sometimes I wouldn't. At lunchtime, I'd have another try. I had an old flat Wolseley handpiece; we used to call it the brick, cos it was a big heavy thing, big long flat thing. Of course, they had Coopers or Listers and they were a real streamlined handpiece . . . I always wanted to get the sheep opened up so I could get on to the long blow – felt like you were doing something then, but it used to take me a while – I've seen smoke come off the handpiece. I tried to shear a ram one day and the handpiece got that hot, I had to stop. So I never attacked a ram again; always picked out a decent sheep to shear and learned that way.'

He remembers those days fondly, recalling his mother used to bring smoko down to the shed:

'She'd make the smokos and we'd take it over with us in the morning; and then she would bring the smoko over for the afternoon when she came with lunch. We had no electric jug, and we used to boil the kettle with the blowlamp for the tractor. We had a blowlamp that used to heat the Bulldog tractor up. Why there were no electric jugs I don't know . . . One thing the shearers always liked was a cup of cold tea after smoko; they liked to fill their mug up with tea and put it by their stand and they'd sip away on that in the next two hours.'

Typical Canterbury summer weather meant the land was often parched. 'You could see a mouse run across the paddock from a hundred metres, that's how dry it was. It was dry.' And as for what the shearers were wearing, Murray remembers the uniform well.

'I reckon if they took their pants off some of them could just about stand up by themselves – that much grease on them – and they'd have a black singlet on . . . Godfrey Bowen started that idea, I think. And cos I used to put their front leg underneath my arm I used to like to have a longer sleeve on my jersey, so their hoof wasn't into my arm. Some of the old shearers all did that – put one of the legs, the off leg, underneath their arm to hold the sheep there and then take the belly that way. But today they just sit it out and down and take the belly off with the two legs free.'

Murray eventually moved to Hororata and took over the sawmill there for a time, and then owned the Hororata store and carted hay over the summer. He filled in at Terrace Station a number of times – even being asked by Godfrey Hall, the station owner at the time, to shear his rams.

'I can remember when I was doing Godfrey's, but it would have been a few years later that I'd gone in there to shear and Godfrey come up to me and said, "Murray, I've bought some new Corriedale rams. I want you to shear them." I stuck me chest out. He'd bought these new Corriedale rams, and wanted me to shear them.'

He remembers at one stage shearing alongside Ian Taylor and gun shearer Ray Pirika. Ray tried to convince Murray to change the way he took the belly off.

'Ray Pirika was a big man, big fella, you know, an ex-wrestler, real nice; Chatham Islanders were nice people. He said, "You do most things right, but you take the belly off the old style." But I couldn't change – if I'd been going out shearing all the time I probably would have.'

While the years may have slipped past and there have been many changes in the district, particularly with the invasion of dairy cows, the Terrace Station woolshed has remained a constant presence.

'I'll never forget it. I still look at it today, you know – and it still looks as big and ugly as it did when I first went to it. But it would have been great in the old days cos it was all blades, and you'd walk in there with twenty shearers and it would be that quiet you'd just hear *click, click*.'

Those standing in the breach to aid the industry had other plans for their futures than a long-term career in the sheds. But a couple who have loved the sheds for decades and have channelled boundless energy into the shearing industry are Peter and Elsie Lyon. Their name is synonymous with quality shearing, and their work ensures Peter Lyon Shearing is the leader in the competitive contracting scene. The story of how they grew the business into the multi-million dollar run it is today is a tale of seized opportunities, grit and determination, and oodles of hard work.

Peter and Elsie Lyon – 'A little old shearing business under the trees'

'We were in the Mossburn Hotel and we were working for Fred, and he said to us, "I'm thinking about getting out of my run; do you want to buy it?" And with no planning, no nothing, we said, "Oh, suppose so." You know, "Do you want to take it over" or whatever the words are . . . a big business meeting in the Mossburn pub over a raspberry and Coke.'

There's a bit of a grin on Peter's face as he shares that memory. A lot of water has flowed under many bridges since that simple verbal agreement between Fred Wybrow and Elsie and Peter Lyon. Peter Lyon Shearing, based in Alexandra, now employs 240 people at the peak of the season, shearing the equivalent of two million sheep annually. The infrastructure they own to support such an undertaking is on a scale to match. Peter and Elsie are stalwarts of the shearing and wool industries.

They are the biggest merino contractors in the country, and both are life members of the New Zealand Merino Shearing Society; Elsie has successfully competed at the highest level in woolhandling, coming third in the 1988 world championship and winning the New Zealand championships in 1996. Peter has been recognised as a Master Shearer and a life member of the New Zealand Shearing Contractors Association (of which he was president for ten years), and he has successfully competed at the highest levels in New Zealand. He also competed in Australia, as well as being part of a four-stand world record in 1982. But how did a young woman from rural Hawkes Bay and a young man from Pleasant Point, South Canterbury join forces and together become one of the powerhouses of New Zealand contract shearing?

Elsie –

As Parehuia Whakaangi Elsie Lyon (née Karekare) speaks, she transports the listener back to another time and place, away from her and Peter's large, two-storey home in Alexandra to 1960s Pakipaki, a small Māori village of around 300 with four marae, just south of Hastings.

'My dad's family, they had a marae that their family all actually lived in when he was a child. And then the other families had their marae that they all lived in. There was no such thing as a house each; and then, I suppose, as they got older and more independent, they all built their own. But the land was always there, and so they just had to build on it. In our little corner, there was my grandmother's house, and then Dad was next door and my aunty was next door to that, and his other, youngest brother was next door to that; so in the four houses we were all in one big family there – it's just how it was.'

With warmth and a sense of nostalgia Elsie shares memories of her years growing up amongst whānau. She talks with pleasure of having a 'really good upbringing' with her hard-working parents, Te Here Lillian Karekare and Ketia Roy Karekare, and two siblings, Henare and Roylene.

Born in 1963, Elsie reflects that hers was a childhood with no expectation of being wrapped in cotton-wool like the children of today. The village children made their own fun: swimming in the creek, taking turns on the one bicycle they all shared, swinging on the rope swing, collecting and selling bottles (the words of her older siblings still ring in her ears: 'be careful', 'don't break them', 'did you clean yours?'). There were trips to the beach with her father: 'We were really lucky we were allowed to camp there – he would light the fire and he would cook us fish in the morning. It was really lovely.' Her dad would dive for kina, pāua and crays, and the children would collect rock snails or booboos to cook and eat. Parties, too, were a part of life: the rattle of crates would be heard, the guitar would come out and family and friends would gather, although Elsie remembers being not so thrilled if her proud dad dragged her out of bed to show off her Māori breadmaking ability.

Thursday night was always shopping night: the family would drive in their Vauxhall Velox to Havelock North, where her father would frequent the pub while the children and Elsie's mum would do the shopping. After the shopping was completed the routine was for their mother to meet up with her husband for a drink.

'And then us kids would be sitting in the car outside the pub and we'd get a pie and some chips for tea and a bottle of Coke. We'd sit there and wait till they had finished, and then we'd go home. And that was every Thursday night, pretty much. Everyone seemed really happy, it didn't seem angry . . . and we didn't feel unsafe; we felt quite safe just sitting there. You wouldn't be allowed to now. And that was always the thing: we never thought anything of it, but I probably wouldn't have done it with my kids. But I think our worlds have changed a wee bit.

'Our community life . . . we'd have what they would call gala days at the pā or marae, and that would happen maybe once a month. Somebody would be fund-raising for the church or a family or something. So we used to have raffles, they'd play games for us kids, and then they'd have horse rides. My dad's sister, she used to make Māori bread or rēwena and we'd raffle it off, and they'd have baking stalls where we would sell. They were real good times. And then we would have Bridge Pa, I suppose that's only ten k down the road, then Fernhill. All the marae communities would support each other, and so if somebody used to make a really beautiful sponge they'd all rush and get those, or the rēwena.'

There were household chores to do; everyone chipped in, and by the age of twelve the children had to take turns to cook tea. 'Dad used to make sure we knew how to cook.' Manners were enforced, and Elsie laughs at the thought of her own children asking to be excused from the table after a meal, like she was made to. A strong work ethic was ingrained in Elsie at a young age: 'If you needed something, you had to go out and work for it. There was never anything just given or bought; you had to save and you had to buy your own things.' Aged around eight, she would head out at weekends with her aunty to pick vegetables. The grower would come around to pick up the workers from the village and they would all pile on the back of the flatdeck truck and head out for the day.

Elsie's earliest recollection of the shed is from around age four:

'My mum used to cook for the shearers every now and then . . . it was our uncle who had the business, or he used to do the run and so Mum used to go out and help him and do the cooking. I can remember being the little kid in the shearers' quarters, and I'd get chocolates and fizzy and things from all the shearers. I can also remember going and playing with the cocky's kids, or the farmer's kids. Their mum used to make beautiful pikelets and things, and so I would be there a lot of the time in the day playing with the kids, and then get home at night.

'I was probably about fourteen when I first went out, cos Mum was working in the sheds by then, just doing a little bit here and there. It was always a family gang . . . we were all picked up from Paki Paki, so it was everyone there that went. Learning how to rousie was really hard work, but I can remember running around all day and then heading home at night, I was so tired. And my cousins teasing me, the shearers, telling you to pick up, hold this, do that, and then they'd all laugh at you and I wouldn't even know what they were laughing at. I remember I ended up going out with a lady called Gert Hokianga; she was an old legend in the sheds in our day. Her sons and nephews all used to work in the sheds as well, so if I ended up in her gang she would always look after me cos I was a hard worker, so I probably used to do a bit of her work but I didn't know it. She was so funny. She'd have housie on a certain night, so she'd say to the gang, "Finishing at four o'clock . . . I've got my housie", and always they'd finish; she was like the boss. She'd work with one hand on her broom – she never used two hands to rousie – and they'd never do anything; if she wasn't finished working around them, they'd wait. They'd never step over cos she'd just get her broom and whack them. She was so funny . . . they'd be her nephew or her son or whoever: she didn't care, she'd just whack them and they'd wait. They all knew to wait. She was always good with me and she'd say, "Send that girl out with me again, I like her, she's a good hard worker," so I'd always end up with her – so it was good cos on certain days you'd get home early cos it was her housie night.'

Schooling was, of course, a feature of life. Elsie attended Paki Paki Primary, Hastings Intermediate and Hastings Girls' High School. While she was no slouch at school, she left just before completing School Certificate and headed to the sheds.

'Usually I would be with Mum cos I was still learning . . . She was a little bit growly at times, and so they'd all be moaning about her growling and I'd be running around all day trying to keep everyone happy. I think I just picked up skills from that. I can remember one of my cousins showing me how to skirt the wool and make sure that you can see it all properly as it's coming towards you, and I can relate to that still now if I'm teaching someone how to do that. Way back when I was fourteen, I still remember her showing me how to stand side-on to a table and look at what was coming towards you, pulling it up, shaking it off . . . so that's how I teach people today.'

Through the shearing season, aside from a couple of stay-outs, it was a five o'clock wake-up with a quick brekkie, lunch made by her dad, then climbing into the smoke-filled van; 'everyone smoked' and it was a matter of 'scrambling around in the dark trying to find a seat and then head off to the next place, and if they weren't awake you'd have to sit in the van'.

'If you knew them quite well you'd get sent out to wake them up, and then you'd head off. There'd be talking and laughing, and if somebody was hungry they'd say, "What did Uncle Roy make you for lunch?" "Oh, I don't know" – and they'd say, "Pass me your bag" – so they're starting to eat your lunch before you're even getting to work. Dad was quite clever: he'd know that that could sometimes happen, so he'd make quite a lot and always make sure I had stuff left behind for me.'

After a couple of seasons Elsie was enticed by a girlfriend to do a pre-lamb shear down south. Warned she would need plenty of warm clothes, Elsie headed down for a six-week stint with Potae Shearing, based in Milton. 'That was a whole different thing for me because it was a big operation like this.' Her friend knew a lot of people, but Elsie didn't know anyone. 'I said, "They won't split us up will they? They know that we're going to work together," cos I was getting nervous, thinking, "Oh my God."' After ten days of working together the two girls were split up and Elsie made her way in a new gang with the help of a nice cook, a helpful wool classer and a good ganger.

After the North Island main shear Elsie came back and spent the summer working for Terry Burling at Middlemarch. The following season found her and another girlfriend flatting in Invercargill, working for contractor Alistair Shirley. Atrocious weather was playing havoc with work.

'It was in January, summertime, with another girlfriend, and we ended up flatting down there. I hated it. I hated Invercargill, cos it rained all the time. I'd never seen so much rain and we were flatting, and you didn't get much work, so how are you going to pay your bills and stuff like that. And then the boss, he said, "Do you want to go up Central?" This was in July. He said, "Do you want to go up Central for a couple, three weeks' work?" Cos a couple of the shearers were going – a couple of the older guys – and they said they'd give us a ride up, me and my girlfriend. And we said, "I suppose it beats sitting around here." So we ended up coming up and working for Vic Harrex, over the road; that was his house that we ended up owning ourselves. We did four or five weeks up here and I loved it – loved the merinos and the weather – and I didn't want to go back, and the boss was ringing from Invercargill, "Are you coming? I told you you could go, but you had to come back," and I said, "We don't want to come back."'

Elsie laughs at the recollection. The Central work was the beginning of her love affair with merino.

'It's definitely easier to work, easier on you. We all love it: anyone that's worked with merino wool loves it cos it's softer and not as greasy on you. Normally, if you work at a cross-bred shed, your clothes are covered in black and you feel greasy and itchy, but not the merino wool – it's beautiful.'

Once Vic's stint was finished, Elsie got work on Fred Wybrow's run.

'They had a lot of wether shearing in September, October, start of November, which we don't have a lot of now. They had 10,000 wethers at Coronet Peak and Glencoe Station, a good month's work; so we did that, and then we went home in November. And I said, "Do you think we could come back in the summer?" And he said, "Yeah that'll be good." So we came back in January and pretty much worked right through cos they had a run in Clydevale and Mossburn. We'd end up down there in the summer and up here in the winter, which worked out quite good. It must have been the next year that I met Peter at Coronet Peak, which must have been '82, '83 . . . so I ended up staying here in the end.'

Little did Elsie know then that the territory she was travelling through would become her new home.

'We were coming through Alexandra and I can remember thinking, "Look at all those rocks; what're all those rocks on those hills and how do the sheep live? Do they have sheep on those hills?" And then I ended up living here all those years later. It's funny how it all works out.'

Peter –

'Those days we were a good-living country family, so you built a whole lot of pride into you as a family – not that you got big-headed about it, but we were still the Lyons and our place was called Riverview and it was just a 500-acre farm, but Dad was a good farmer, and we did the right things; we went to church on Sundays. We lived a pretty Presbyterian life in those days, we were brought up pretty correct. Little things, like we weren't allowed to go into our sisters' bedrooms – rules like that are not heard of today. Our church was in the village of Pleasant Point. And rugby club: right from an early age we'd be taken there on Saturday mornings to play in the thirteenth grade; that was when you started, not sure what age that would be – six, seven, eight? Your community was everything, and that's where you built up your reputation, whether it was good or bad. If you were bad in a small area you were history. There was no tolerance: we only had one choice and that was to be good and not let your family down and not let your area down . . . so we had a pretty narrow upbringing.'

The backdrop to Peter's life reflects a very different world to that of Elsie's. His parents, Catherine and Douglas Lyon, raised him and his siblings, Denise, Joy and his twin Gary (always known as Bill) within a strong Presbyterian tradition on their farm at Kākahu, just out of Pleasant Point. Born in 1952, Peter's earliest memories revolve around the farm; getting his first pony at age eight stands out:

'I would get home from school and get on my pony, and I rode around and looked round the sheep for Dad. So, from that time on you're learning things: you learned stock sense, you learned that animals have to be cared for, they've got to drink and eat, and they can't be shut up and all those things; but we just took it all for granted, I guess.'

Peter, too, was no slouch at school and was always good at sport, particularly rugby and running, as well as being a capable equestrian. He spent a year at Timaru Boys' High School as a boarder, where he remembers chasing Dick Tayler (later a champion runner) around the block on their compulsory morning runs. He finished his schooling at Pleasant Point District High School; halfway through his fourth year of high school a shoulder injury curtailed his sporting activity, at which point he decided there was little point in staying.

Keen to purchase his own farm, Peter was never afraid of hard work. Working for his father, eager to earn an extra dollar, he moonlighted for other farmers.

'I'd get up early in the morning. I could always get up good and early – at home I had to make time to go out and help the neighbours, so I'd be out on the tractor at four o'clock and I'd do two, three hours, then come in and Mum would have some breakfast ready, and I'd dash up to the neighbour to start in the woolshed at half past seven. And then you might do another hour at night.'

When it came to taking on shearing, he already knew how to work. Raised on a mixed farm with 1800 ewes, work in the two-stand woolshed was always going to be a feature of Peter's life. The shearers would come in each year around the twentieth of November and the ewes were shorn.

'It would take a whole week to shear those 1800 ewes. The guys would come in and have breakfast with Mum and Dad at the kitchen table, and then the farmers did their own wool; the neighbour would come in and help, and that's the way it was. We would do our dagging: in those days you'd nip a bit of shit off them with the blades, so we didn't put them over the board like they're probably done today.'

When the brothers were considered strong enough, they graduated to pressing, stomping the wool down in the press.

'On the double wooden press with the two clickers on each, you did one each side and you'd be pummelling away at that press. It was pretty hard work when you were young, but I was fit and strong; and if you weren't strong you had a go anyway. Our dads, they weren't shy to let us work, like I probably wouldn't ask my kids to do. My kids are good, they'd do anything, but they wouldn't have been asked to do the things we were at fifteen. We'd be scared that we were going to ruin their backs or whatever. But it wasn't thought of in those days: you were there, you were a unit and you just did the work.'

Peter began to wield a handpiece at age twelve, when he and his brother helped their father with crutching.

'Us boys and Dad would crutch the lambs; they are quite small and manageable at that age, so the biggest thing to manage was the handpiece cos it's got a dangerous bit of gear at the end of it; that's probably your biggest worry. Dad taught us to crutch, and we probably did that for three or four years, and then he always used to shear the odd one that was left over or something he'd got from the sale, so we'd have a bit of a practice with that. Dad could shear, so he could tidy up if we weren't going that good. So, we could basically get the wool off when I went to shearing school; though I'd never done a full day's shearing, or for that matter crutching, I knew the basic fundamentals of holding a handpiece and setting the gear up.

'Going to shearing school was a real learning curve, but I was ready to learn. You've got a bit of a pre-starter when you've done the odd one and you know what a handpiece feels like. So, off to the shearing school. That was really good. What actually caused me to go shearing . . . I can remember sitting on these wool bales and Dad was paying these shearers; one of the shearers was only there three or four days – it was inside a week, anyway – and he earned $144 and I earned $72 for the month. And I said to Dad, "Gee, those jokers earn a bit of dough." And he says, "Yes, son, you ought to go out and get a bit of that." So we got booked into a Wool Board shearing course out at Rangitātā Island, which is where all South Canterbury boys learned to shear, cos I can still see the place out at Slee and Meyers at Rangitātā Island. So, I did this course, and it was a fella called Robin Kidd – Robin was another South Canterbury boy. And he said to me at the end of it, "What are you doing, Peter?" And I said, "Oh, working at home for Dad, and I go out and do a bit." "You want to have a go at this," he said, "you've got a bit of potential." That's what Robin said to me, so I went home and then must have got myself a job and away I went shearing. That was back somewhere in the sixties.'

His first shearing job at age nineteen, he thinks, was teed up by Robin. All the stands were full at Pleasant Point, so Peter worked for Temuka shearer Frank Jones, or Jake as he was known. Peter remembers it being physically hard work, but recalls, 'I always enjoyed it, so that helps you get through the hard days.' The focus was very much on the quality of the shear and not the numbers.

'Those days we were still really conscious of doing a good job, and the only way you can do a good job when you're learning is to go slow. So, we're doing a hundred, 120 a day.'

While the first 100 was no problem, reaching his first 200 was more of a battle.

'Oh, I was pretty much there straight away you know; a hundred is only twelve an hour or so, or twenty-five a run, we say. So once you could shear a hundred your next aim was to shear two hundred and probably, to be honest, it took me a couple of years to get up there; but at the same time we had it drummed into us, get it right – you've got to have your job right. Sometimes the young lads today are a bit the other way: they try and go too fast so they can get a stand, cos they think they've got to do a number. We got a stand because we were keen and we were doing a good job; nowadays they think they'll get a stand if they shear fast, so we've got a bit of the opposite today – we've got to knock them back.'

Peter thinks he worked three or four seasons for Jake. When the sheep were dry he worked Monday to Saturday dinnertime.

'It was like a community thing again, and Jake was good. He was a Roman Catholic – he wasn't a nutty religious fella, but he was a Catholic, so he had church on Sunday and often rugby to watch on Saturday afternoon. And that's a fact – like, that was it: you either had to knock off to play your rugby or go and watch it, and that's the way it was all round South Canterbury, Temuka, Geraldine, Pleasant Point, Fairlie. You had to be at that park to watch your local rugby team.'

Peter honed his skills working for Jake, shearing cross-breds, the finer-wool Corriedales and merinos. He went on to work for Steve Ryan for a time, getting caught in a shearers' strike in the 1970s; Peter smiles as he talks of it being the only stand he ever lost. Shearing competitions by this time had entered Peter's frame, and he and his good mate John McGillen attended the Waimate Shears together. After losing his stand with Steve Ryan, Peter moved to another contractor, Johnny Walsh. Johnny and Kevin Walsh, as Peter recalls, 'were two really good shearers in South Canterbury':

'That was quite a good era; we could shear, we were winning at the shows – people like Kevin Walsh, John Walsh, John McGillen and Adrian Cox and myself, we were all taught through that same shearing school, through Robin Kidd, and we all would have got national titles. We were pretty strong; there was always someone better right in the top grade, like Roger Cox or Alan Donaldson or some of those top, top guys; but we were right in there, up their backsides.'

Throughout these years Peter had also bought a small farm, married and had two children. He juggled shearing with farming his 1000 ewes on the 240-acre farm. Weekends, from dinnertime Saturday, and any free afternoons were spent on the farm. Where necessary he would get a hand from his father or rope in a mate. Somewhere in there as well he still managed to compete.

'We'd go from blimmin' Invercargill to Masterton . . . not as much as they do today, but we covered the same area; they do twice as many shows today. But we were on the bones of our backsides – every dollar was spent and we had to keep working.'

One aspect of the competitions was the downtime afterwards.

'In those days . . . when the competition was on it was usually at the end of a period, so the guys had a pocketful of money and they were wanting to relax a bit. They'd done the hard yards, so afterwards it was always generally a good social night. We would get a little bit stupid – nothing too silly, we all had partners – and we used to have a bit of fun, but we never did any damage; just a good old-fashioned piss-up where everybody would enjoy everybody's company, and that's the way it was. You know, tell everybody how good they were; the girls were there and they'd be having their fun, and there'd always be someone trying to pash somebody they shouldn't. I suppose it was always a bit of fun.'

But when his marriage ended, Peter had to sell half the farm. Shortly after, he headed to the North Island.

'I did things arse about face. I did things when I was thirty that I should have done when I was twenty.'

He thinks it was 1981; by then he had won the Golden Shears Intermediate grade in 1974 and the Senior grade in 1975 and was competing at the Open level, coming fifth in 1978 and 1979. Rick Pivac, a mate and contractor in the Wairarapa he competed against and drank with after the shows, took him on. With Pirinoa and Te Mai stations in Rick's run, there were plenty of sheep to shear. Peter laughs at the memory of settling into the gang:

'I was the only white man in the gang and we were in up to seven-stand sheds, which means fifteen of us, so you soon learned to tone in.'

This 'little white boy' from South Canterbury was treated with a degree of suspicion initially:

'The guys in the gang thought I was a private detective sent in amongst them . . . I was white, I had short hair, I had a brand new car, a little Datsun 180B, I had my own grinder, and they wondered what this fella was doing amongst them.'

Peter enjoyed the hard physical graft in Rick's gangs.

'My best shearing was in the shed. Rick was one guy I could never beat, he was phenomenal, strong, unorthodox and quick. I could handle most others that came my way. My motto was "shear quick and clean", and I was proud of that.

'And that was over eight hours – just strong and to a reasonable standard, never got chipped or anything. You're always proud of that. And would I be shearing two million sheep today or the equivalent if I didn't have those standards? It's not really likely, is it? You can't just be a businessman and run a shearing business; that's not the way shearers let you operate, I'm afraid.'

Peter worked his second season in the North Island for Bill Morrison, who ran everything with military precision.

'Bill could run his place almost like the army cos he had pretty high expectations of himself; also, I'd shorn with Bill in the Senior final that I won, so he was another mate through shearing. I did all the rest of my shearing in the North Island with Bill.'

The north–south migration, common to many shearers, became Peter's regular pattern, too.

'You were in the North Island always from November/December . . . and then again you'd go there in very late February or you'd go up and shear at the Golden Shears; in those days the competition season went from spring to autumn, so you were at the end of the season. And you would stay up north and do March and April and then head back down south, do a bit of crutching and then your pre-lamb. That's how the season went.'

Throughout this time Peter was free to just get on and shear, and he revelled in the opportunity, shearing around 65,000 sheep a year. The impact of the experience was twofold.

'Once I'd been there I didn't want to go back home into what I call cocky shearing, where farmers were taking our wool away and you'd stop and have a bit of a yarn to them (or you were expected to). I improved my shearing probably by twenty a run by just that contract environment, where everybody was paid for what they do, and you were totally there in the industry professionally.'

Throughout this time Peter continued to do well in competitions.

'My biggest thrill was – I was just short of the best, but I shore with the best in those days. It was a pretty good effort: I made a couple of Golden Shears Open finals, which is the last six. I ran plenty of seconds to the likes of Roger Cox, Alan Donaldson and Colin King in the earlier days. I had quite bit of success around Canterbury: I'd win half a dozen shows around there and then I'd always be in the major final. I made the Caltex Open Final probably eleven times (it's called the Wrightson's National now; it's the multi-breed event); I'd always be in the first two or three in the multi-breed. They used to have a South Island All Breeds title and I was good enough to win that, but in my era there was always somebody . . . I was good enough to get competition cheeky, meaning I could be within a point or half a point, but they were just too good. But when I look back, my best work was always done in the shed, and I was conscious I had to earn money. Today, these professional guys will take two days off before a show, but we worked right up to the night before and then travelled overnight and then we were shearing the next day.

'We all tended to get on. I suppose we had our opinions about each other, but we got on. You didn't go round hating the opposition or anything, cos you were probably working with them the next day; and yet it was very competitive. But you've got shearing gear and it's very specialist, so the big thing was, the better you were with the gear, the less you showed people. That was the secret in our day: the better you could do your gear, the more chance you'd give yourself.'

He laughs when asked if competitors kept their knowledge to themselves:

'Especially in those days; especially when we started. Like, Snow Quinn: he's who we used to idolise. Obviously, David's [Fagan] come along, but in my day one of our best competitors was Roger Cox, and he's a great guy. But they're always playing with tactics . . . you're trying to beat them and they're trying to screw with your head. They might even show you a comb that they're not even using . . . to be fair, once it got to David's era things changed a bit – it got more open. Guys share their knowledge better, and they back themselves . . . If guys thought they had something up their sleeve they'd keep it there, and that's just the way we were. I always looked at the guys that were better than me and thought, "Just show me what you're using, you prick." So there was always a lot of gamesmanship, and because there's so much going on it's quite easy to get at somebody one way or another, whether you're physically – like Snow, he was a tall man with long arms, and might say to you, "They're pretty big, these sheep" . . . you just know he's big, he's got the reach, he's going to be better than you.

'I remember in the Golden Shears, I'd made my first Open semi-final and was walking up on the stage behind a fella called Tom Brough. I looked up to Tom; he was a bit older than me, but he'd been there for years . . . It's quite steep stairs at Masterton, and he looked around behind and said to me, "You'll be a bit nervous about this, won't you, son." And I'm thinking, he'd know we're up and coming, and gee I was thinking, "I'm not half as nervous as you're going to be in a minute, Tom." But that's what they used to do to us, Tom, Snow . . .'

Peter enjoyed competing, but eventually life took him down quite a different path.

'I was going north and south and actually got in the New Zealand team. It was the time of the wide gear issue in Australia, and I was the team captain that year . . . so we were to go to Australia and we were competing with the narrow gear in union country. And obviously we were wide gear over here, so we were worried that we would be seen as examples. There was already shots fired at a couple of fellas in Australia just before, in that winter, so the Golden Shears Committee organised for us to get stands inside a six-stand gang at Coronet Peak. There was Alan Donaldson, Coxsy and me; Ricky Pivac was the other one, but he couldn't quite come down cos he had his run up there. So we got this job at Coronet Peak and this was our practice three weeks before we went . . . We ended up over at Coronet Peak camped up under the mountain there, shearing these 10,000 wethers, and that's when Elsie – my wife now, but she was working there as one of his shed hands – that's where we met, in 1983 . . . Obviously, we went to Australia and competed and came back, and I went backwards and forwards; but I kept in touch with Elsie, and then six months later we crossed each other's paths and got together. We did this north–south thing and I ended up working for her boss, Fred Wybrow.'

Peter Lyon Shearing –

At the 'big' business meeting with Fred a plan for the takeover was discussed, and within the next few months it was implemented. As Peter recalls, the plan was simple:

'We said yes. We said, "What's the go?" And he said, "I want to stay in it till after pre-lamb," and that was okay. And he said, "But if you work pre-lamb I'll introduce you to all the farmers and move you about – maybe not where you want to be sometimes, but you will meet everybody." And that's what I did. Then, that October – it was 1985 – we did our first shed, Coronet Peak. So, it was Elsie's connection with Fred. She was a very good woolhandler; she did a lot of grading of the wool, so she was well thought of among the clients, and it was only a small run, like we had three gangs at the peak . . . one six-stand gang and a couple of three-stands – and that was it when we took over.'

Peter acknowledges Bill Morrison as a good friend who enabled them to set out on this new venture, acting not only as their contracting mentor but also as their bank. Bill, along with Peter's father, loaned some of the money the young couple needed to get established. The nucleus of the run included Coronet Peak, Glencoe, and work up in Ōmarama and Poolburn, but also stretched into Mossburn and Clydesdale, averaging roughly

300,000 sheep. With enthusiasm and age on their side – Peter was thirty-two and Elsie twenty-one – they got stuck in. Elsie laughs at their naivety:

'Fred had said he'd only wanted to do it for five years, and that's pretty much all you had to do – just do five years and then get your money and get out – and that's what his attitude was. We thought that sounded alright . . . At that stage I'd decided, I want to be out of it before I have children, so thirty-two years on here we are: two adult children, still going.'

Very quickly after this purchase, within the first six months, the young contractors picked up two other runs: that of Jenny Streets (Jenny Te Whata), which was Murray McSkimming's old run and Graeme Bell's.

For Elsie there are two significant staff who came across from those runs and were to make a positive impact on how Peter and she were able to work. June Heatherington from the Te Whata run 'was *amazing* – she was great with the cooking, the people and me. She used to come and help me with [my daughter] Juliette. Oh God, I don't know what I would have done without her.' And Clyde Oliver, Graeme Bell's father-in-law, came across as their handyman. Elsie remembers him knowing all the farmers and that he was a 'real good old guy'. A lasting friendship developed with both.

As if the challenges of establishing themselves as contractors were not big enough, Elsie and Peter picked up another three small runs, managing to amalgamate six runs within eighteen months from their first purchase. As Peter remembers it, there was a reason behind the readiness for the sales. 'What was actually worrying these fellas was GST had just come in, and they were scared of it, or so they said. It was a big thing. And of course, we were young, we were certainly keen, so we didn't give a toss . . .'

While Peter and Elsie may have earned their stripes in the shed and competitively, it's fair to say there was much to learn when it came to running their new business. Peter reflects on the balancing of various roles:

'Initially, of course, you think you can do everything, and you think you're not working unless you're physically shearing, so that's the biggest thing to get your head around there. And you're used to getting paid for what you do, so what you do is shear sheep and that's what you think you get paid the most money from.

A Peter Lyon Shearing gang working hard at Mt Nicholas Station. Laughter is an important ingredient in the fast-paced, demanding shed environment.

When we started off we had a couple of gangs going initially, then we built up to three or four at Christmas time. Then I tried to go out and do a day's work; I might take the short day. And with just the two of us running it, I was involved with much more of the accounts and the office side of it as well . . . We did our own GST, which was done more or less by hand then. If you go back to 1985 I think we actually developed our own spreadsheet with another chap to do all that.

'So, initially I tried to get out shearing, and . . . it was pretty good money. Elsie would often do three jobs – she'd cook, rousie and organise as well. We were pretty productive, and you end up with what you think is good profits. But as you get bigger, you know it's a false economy and that eight hours is a long time to spend in a woolshed when you're trying to run a business. Because actually you're in the shed for ten hours cos you've got two half-hour breaks and an hour for lunch, plus possibly an hour each way driving, so you're twelve hours away from home. But you're a shearer and not a businessman, so you've just got another thinking cap on, or you have to change roles, and that's probably the hardest thing. I still think today that I've got the easy job; being the boss has its difficulties, but it's still not as hard as physically being out there doing it. I've never lost respect for that.'

Elsie remembers that they taught themselves how to do all the administrative side of things: payroll, PAYE, GST. But the role she found most challenging was that of cook. 'It was a bit of a nightmare and I hated it. I was so pleased when we were big enough to hire a cook.' Their days would start somewhere between three-thirty and four in the morning, and they would potentially not finish until ten at night. When Elsie was eight and a half months pregnant with daughter Juliette in 1989, she found herself at pre-lamb, their busiest time, cooking and baking to fill eight tucker boxes each day. It is little wonder that she ended up in Clyde Hospital with high blood pressure. Elsie laughs at the manic situation:

'The nurse was going nuts, the midwife was going nuts – "Every time you come in to see her, you bring some more work for her! She needs rest." And my blood pressure hadn't gone down at all after seven days, so, "We were hoping you were going to have the baby here, but you need to go down to Dunedin." So, I ended up heading down to Dunedin and she was born the twenty-seventh of July. We were busy, obviously, with eight gangs, and Peter couldn't come, and I ended up getting a ride down to Dunedin with some man from the hospital here – some strange guy.'

Peter first saw his daughter two days after she was born, when he brought Elsie and Juliette home.

On her return, Elsie was straight back into work, supported initially by her mother and a girlfriend, and later by June. Juliette became very used to being 'thrown' into her car seat and taken out to sheds. After having Juliette in their busiest time, Elsie was determined to do things differently next time. 'I thought to myself, never again, that was just so hard . . . I'll just have to plan my next pregnancy.' She can laugh about it now, but at the time, she wasn't overjoyed to discover that their second baby was due at the second-busiest time of the year – Christmas. Nathan was born on Christmas Day 1991. Elsie jokes that they stopped at two children because they couldn't chance having another one at a stupid time. In a moment of reflection, she adds that they are 'really lucky the kids are as good as they are, because sometimes we were so busy they just had to go with the flow'.

When she needed to, Elsie would still work in the shed.

'I have to admit I've always loved working in the shed; always have. Right from day one, right from when I was fourteen and went out to learn, even though I didn't know what I was doing – I was running around like an idiot – I still really enjoyed it. I think it's always been the challenge. I challenge myself every time: if I go into a shed I like to think that we can have it running nice and smooth, everyone working well, everyone happy with what they are doing. And then I challenge myself, if I'm on the broom, to get every belly, get every crutch, keep it all tidy, help those girls on the table. That's what's in my head. And when I get on the pickup, try to land my fleeces nice and be able to keep the wool flow running. And I try to encourage the people I work around to have it in their head, too.

'I remember working with a girl on a closed board . . . I'd do my broom work and she'd follow me, and we'd go around again. Then at smoko she was sitting all by herself, and I said, "Why are you sitting here all by yourself?" And she said, "I just get so disappointed in myself." And I say, "Why?" And she said, "Because I don't know how you can keep that board all nice and clean and tidy and everything in its place, and every time I get on there I can't: it's a mess and I just want to cry." And I said, "What do you mean?" And she said, "Well you watch me, I get in a real . . ." I said, "That's because you're trying to do everything physically without using your head. You know for a fact that if that one's coming off and this one's coming off, which one's more important. And timing: it's all about your timing." And so I said, "I'll go on the broom first and start and show

you, and then you look at what I do and think about how you do it." By the end of
the day she had it down pat, and she was so proud of herself and so happy. And
she wasn't a learner: she'd done it for a couple of years, and that's why she was
so disappointed in herself – she couldn't keep it as tidy as it should have been. I
said, "Look at all those piles of wool everywhere, that's just getting in your head
because it looks untidy. Put it in something; don't leave it to build up till all of a
sudden you don't know where you are and what you're doing." She got it. And I
said, "You've got to work hard at it and get your head right and do it." You have to
keep yourself motivated, and a lot of them think it's just physical, but you actually
have to use your head or it doesn't work.'

Peter and Elsie have run Peter Lyon Shearing for over three decades now
and the business has grown significantly. There is a sense of satisfaction in
what they have achieved. As he elaborates on the business's development,
the last thing Peter wants is to be 'an arrogant prick – a Donald Trump'.
There is no room for skiting, but he answers questions as straight as he can:

'I think we actually pass about 2.8 million sheep through our legs . . . I suppose at
the end of the day we shear probably 1.6 million and then crutch another million,
cos we crutch most of our sheep once. Everything that gets shorn pre-lamb is
crutched, and then all the lambs are crutched, so that's their offspring, and then all
the adult sheep are crutched before they go to the lamb. So, it's a little bit unique:
you would find in a run like Ewen's [Mackintosh Shearing] they would hardly do
any crutching – it's a bad word – but it's good business to us. It's a South Island
thing and a high-country thing . . . our crutchings down here are worth more than
full-wool cross-bred fleece, so it's important that we sort it, too. Our crutching
season is very good: we crutch most of March and May and our shearers do it. It's
very good income, and the girls go out and sort the wool cos it's worth $3; merino
dags at the moment are worth $3, so the crutchings are actually worth $6 or $7.
You only get a little bit off each sheep, but they add up; I think they can almost pay
the crutching bill. On a big station up north, if they've got 20,000 ewes it will cost
them $20,000 to dag them up; there's no comeback there. Whereas if we've 20,000
ewes down here they'd get a little bit of a wool cheque back – in the fine wool.'

The pre-lamb is a daunting task each year: for five to six weeks their
shearers shear up to 17,000 sheep a day. It leaves little time to think.

'It's really meaty then, and our summer's quite good too. Obviously the bigger you are in pre-lamb you're still big in the summer, but we still have gaps . . . we have our quiet spots, but not for long.

'Even though it's in Central Otago and people assume we shear more fine wool, we actually don't. We still shear about forty per cent fine wool to sixty per cent cross-bred. And that's because the merino season is only three months, four months of the year, and the cross-bred season rates over twelve months. There's still a lot of cross-breds down here cos some of the merino, ex-merino places have got improved irrigated flats, so they're using the irrigated flats to produce their lambs and using the run-off where they might have used merinos for wool. So it's all to do with tenure review – that's a big issue.* The tenure review has taken the top country, so they've used the money that they've got for that to develop their low country, and that's often led to a bit of a change in sheep breed.

'Lyon Shearing goes from coast to mountains, so you've got a good geographical spread. That gives you climate, that gives you land change; and because you're running east to west it gives you a lot of seasonal change. The first of spring at Glenorchy you might say is in October, first of October, whereas in Palmerston [in North Otago] it's maybe some time in July . . . so we've got the going pretty good, spreading our work east to west . . . And you'll find that most jokers with decent runs – the likes of Barry [Pullin] and them – they've got work out at Akaroa and they've got work at Arthur's Pass.

'As a contractor obviously you have some outstanding people working for you, but they've got to have support from the less outstanding people to make the team. As a contractor you might promote some special person that you have working for you, but you still need those other five shearers to make a six-stand gang . . . I've had a lot of shearers work here over the years, and they come and go not because they don't want to be here, but they want to move on; they're doing things with their lives, and we're just a means to an end. I've had David [Fagan], [Johnny] Kirkpatrick, all the guys through the years – Mickey [aka Alan] MacDonald, Digger Balme, Edsel Forde, Dion Morrell, Mana Te Whata, who won the show there five times. But what you have to be prepared for when you lose a good man, you've got to be able to cope. And I've learned someone will step up in his place.'

* The process of the review of Crown pastoral leases in the high country of the South Island through the Crown Pastoral Land Act 1998. The voluntary review enables the freeholding of parts of the leased land while the remainder of the land returns to Crown ownership, often becoming conservation areas. The Labour government has since stopped the continuation of this process.

The couple has battled through 'eye-watering' interest rates in the first years of starting the business, the loss of runs to dairy conversions and tenure review, changes in employment law, the increasing legislation around health and safety, and societal changes and expectations. Peter reflects on their success, and, shooting straight from the hip – as he seems prone to do – he assesses how things have evolved for the couple:

'We're seen as a team. I mean, we're about as different as you can get: one's black, one's white, one's old, one's younger . . . I'm soft, she's hard (if you're both soft, you get walked over; if you're both hard, you have no one). Plus, we know our stuff: we know how to shear a sheep, and Elsie has been in a world championship final in her day – she's good, she enjoys the shed; it's not as if she's judging whether she should be there or what I should be doing.

   'One thing that will stick with me forever is that we do get fantastic loyalty from our clients. But there's a lot going on when you go onto a place to shear a guy's sheep. First of all, he's got to know you're going to be fair with your price, and then he's got to be able to trust the people on the farm – that they're going to care for his stock, they're going to shear the sheep properly, they're going to do the wool good. He [the contractor] is going to feed them properly cos he doesn't want them pissing off during the day to the shop. So there's a lot of reasons they've got to consider before they either get you or get rid of you. It's not just, "Oh, I'll just get someone else" – there's a lot of things to consider. Whether they're going to be there on time; if I want to have a few to dag for the works tomorrow, can he get a couple of guys there? It's all those things.'

Someone who has shorn for them for years is Kelly Hokianga. Treated like family, he has observed them closely over the years and comments on how hard the couple has worked. He laughingly suggests Elsie's 'a little fox terrier with a pitbull attitude – she's a wild little thing'.

   On the surface Peter's and Elsie's childhoods were disparate, but the underlying values of family, community and hard work were instilled in both of them from a young age; this, perhaps, is part of what has unified them and ensured their success. Reflecting on the significance of Peter Lyon Shearing to the industry and to the Alexandra community and economy, Peter is willing to admit that 'it's pretty massive for a little old shearing business under the trees . . .'

Peter and Elsie Lyon recieving their Lifetime Memberships to the New Zealand Merino Shearing Society, 2015. *Otago Daily Times*

When a prime minister can beat the great Sir David Fagan in a shear-off, even if there may have been a bit of good-natured doctoring, you know that shearing's reach is broad. There are many around the country who would have tried their hand divesting a sheep of its wool, but only those with the stoutest of hearts take it on professionally. When the shearing 'bug' is caught, though, it is no segregator of gender, age or race. The issue is not who you are, but what you can do; there is little tolerance for those who don't apply themselves well to the task. Everyone must prove their worth in the shed. The skills needed to become a successful professional shearer take time and dedication to develop. Understanding just what it takes to shear thousands of sheep a year is reflected on in the next section.

The count out pens, a place for farmers to cast their eye over the quality of the shearers' work, Mangaorapa Station.

2

Their Work

While townies are still tucked up in their beds with a good couple of hours to sleep and dream, anywhere there is a sheep to be shorn there will be shearing crews rousing themselves, staggering out of bed bleary-eyed to ready themselves for another day in the sheds. They pile into contractors' vans or drive themselves to their 'office', whether it's fifteen minutes away or more than an hour. No matter the distance, the shearers will pick up their handpiece, pull the cord at 7 am, and shear for eight hours.

For those crews on stay-out, the 'office' may only be metres from their beds. The daily commute is potentially reduced to a few minutes, and the opportunity to catch extra shut-eye snuggled under a duvet or tucked in a sleeping bag is a welcome respite. Although, of course, there's little opportunity for a lie-in if they're working a nine-hour day: for then the cord is pulled at 5 am.

Once in the shed the degree of physical exertion by the shearer can be compared to that of a long-distance runner. Shearing 300 sheep has been likened to running a marathon. If that is indeed the case, many professional shearers could easily be doing the equivalent of at least five marathons a week. Not bad by anyone's reckoning, and little wonder that shearing is considered one of the world's most physically demanding occupations.

The use of the 'top two inches' is essential, too, when grappling with the myriad variables affecting a shearer on any given day. Issues can include anything, such as a badly laid-out shearing shed, antiquated shearing machines or gear that's not going well, sheep in poor condition, stroppy sheep, cotty fleeces, farmers not ensuring their sheep are emptied, a woolhandler not moving the fleece away quickly enough or constantly bumping them . . . it may be freezing cold because it's the middle of winter and in the midst of pre-lamb shearing in Central Otago, or the shed is stinking hot because the roof is low and it's thirty degrees outside in the Mackenzie Country. And that's not even including the sheer physicality of dragging a sheep (which nowadays can be anywhere from sixty to 100 kilos depending on the region and the breed) out of the pen, keeping it in position and decloaking it with a handpiece or a pair of blades that have the ability to maim in a moment of inattention or if a stroppy animal kicks it out of a shearer's hand.

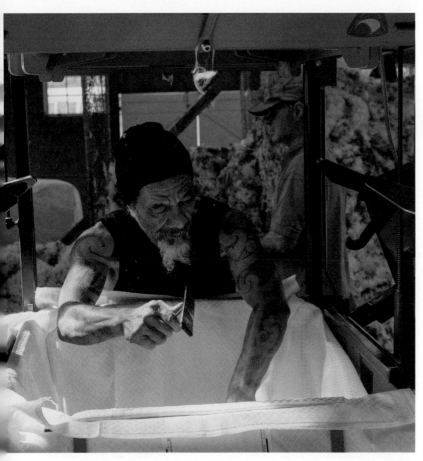

A presser is constantly in motion. It's not just the physical demands of filling and pressing the bales, labelling them and stacking the 180–200kg bales but they must keep the shearers' pens full and are called on to do all manner of tasks to keep the shed running smoothly. In the old days they would also chop up the mutton for the cook.

The fact that a shearer's wage is determined solely by the number of sheep they shear is a great motivator. Those keen to earn well will work hard at perfecting their shearing skills and motor through as many sheep as they can, competing against themselves and the shearers around them. There's a lot of testosterone in a shed, and it is natural in the hard environment for the shearers to pit themselves against each other. Who can get the biggest tally for the day, and then sustain it, is all part of the battle. While the ringers' stand is a thing of the past, the old guns still hold on to the drive to be number one in the shed, and those who are good like to work alongside the best in the business. As Eddie Parkinson bluntly says, 'Good shearers don't mix with useless shearers.'

No matter whether you're working in a shed with 100 sheep in the yard or at a station with a swathe of 20,000 coming in from the paddocks in separate mobs, the sheep all need to be dragged across the board and denuded. It is this work that is elaborated on here. Catherine Mullooly, a young shearer, recounts her journey to becoming a professional.

Catherine Mullooly – The obsession

*'I think you can watch shearing and have no idea
what's actually going on quite easily – like you know
nothing about it until you do it. You wouldn't have
any idea how hard it was until you actually did it.'*

That's the view of twenty-eight-year-old Catherine Mullooly. She is sitting in the garage of her home in Piopio because it's currently the one quiet spot available to talk. The grinder on the bench, the buckets storing shearing gear and a few tools lying around speak to the work of the home's occupants. She begins to reflect on a younger version of herself:

'I was a bit obsessed . . . I was pretty shy and I'd just stand in the doorway. But I remember when they were shearing at home – at Matawai – at the home shed, cos we lived up the hill their noise would travel up, so I'd wake up, jump out of bed and run down to the shed – Dad would already be down there – but I'd just stand in the doorway, and slowly throughout the day end up in a catching pen or something.'

That shy wee girl, standing watching the shearers at the woolshed door, has morphed into a very strong and capable shearer. She has been shearing full-time for around seven years and there is one thing that Catherine makes very clear:

'It sort of annoys me when people will single you out cos you're a girl, or judge you before you've even started work. They don't know you from a bar of soap – they just know that you're a girl and you're standing there with a bunch of other fellas either side of you, so they've already made their mind up a little bit. That annoys me; just quietly, like; it makes you want to bore in a bit when you get people like that . . . But I don't think it has anything to do with the fact that you're a girl; I think it's your size. Cos I think I'm just as strong as a little guy my size – I could be stronger than some guys I've shorn with . . .

   'And some people are annoyingly soft on me, like they don't want you to shear rams and shit like that; it's just annoying, actually. I chose to do this job, I want to do this job; I'm not going to half-arse it. You know the whole job is shearing everything that's put in front of you – you don't get a choice. And I'd never want any of the guys that I work with to think, "Oh shit, we've got all of these rams to shear today and Cath's with us." I'd never want them to think of me like that, like I'm a hindrance to them – "We're not going to get out of here very fast cause Cath's with us" – so I try my hardest all the time.'

With that straight, how did the little girl who inched her way into the woolshed over the course of a day transform herself into a successful shed and competitive shearer who won the Senior grade at Lochearnhead, Scotland in 2017?

   A 'cocky's kid' from Matawai, Gisborne, Catherine was introduced early to her calling, but it was not until she was at a loose end after finishing seventh form at Gisborne Girls' High School that she began to work in the sheds. She would have been eighteen or nineteen, she thinks, and laughs at the memory: 'I was rousieing and I hated it.' She was working for some friends who had a little open run initially, and then worked for contractors

Kev and Donna Williams in Gisborne for a main shear in the summer and then the second shear in June/July; but again she laughs: 'I remember that I didn't even know to take dags out – you had to take dags out of everything. I got growled at for everything.'

Catherine thinks the first time she held a handpiece was with her dad:

'He was shearing a catching pen full of stragglers or muttons or something and he had quiet Romney ewes then – real big, wool all over their ears and everywhere, and really solid. And I said, "Oh, can I have a go?" And I think he said, "How much of it do you want to do?" And I said, "I want to drag it out – I want to start right from the start." He was like, "Oh okay," or probably "Good luck." And it took me ages, and afterwards my legs . . . my whole body was shaking, it was sore. But yeah, a bit of a thrill when you've finished.'

The time-honoured tradition of finishing the last side and keeping an eye out for every opportunity to shear one in the shed started to consume Catherine.

'I was getting pretty keen on it at work; if ever the boys had to stop during a run, go to the toilet or whatever, I'd be like, "Do you want me to shear one?"'

It was going to a Tectra shearing school up at Rangatira Station, near Te Karaka, when she was twenty-one that cemented her direction:

'A guy, Sonny, and Hadfield Smith, he's from Gisborne, they're a shearing family – Dinno Smith is his brother, he's a bit of a wheel – they were the tutors. Dad had said to me, "You know there's schools, you can do schools to learn this." "What, really?" I don't know how I thought people learned to shear. I hadn't really thought I could be a shearer then, I don't think, I just liked doing it. Dad was the one who tracked down the school. And I think he paid for that; he wanted me to do it.

'It was so cool. I remember being so chuffed every day when I'd go home to Dad – just buzzing. I think you do a bit of paperwork on the first day, and you learn how to take a handpiece semi – not fully apart, but semi, crutch . . . and I think you may have shorn your first sheep by the end of the first day. You should probably know a pattern I think by the end of the four days.'

After that, Catherine's head was turned. 'That's when I really started not paying much attention to my job.' Not everything went her way though; there was someone she had to win over:

'The head rousie . . . she was an older lady, Daph was her name. I get along with her like a house on fire now when I see her, but I used to feel like she was quite hard on me; she used to lose it at me, and I didn't know why cos I thought, "You don't tell me why you're losing it" . . . But now I can see working in a shed is stressful when someone's not pulling their weight; you can just flip out. And I remember when I started doing the last side – some of the boys used to set up a handpiece for me at lunchtime and I'd shear five lambs or whatever it was, and they'd spread that tally out for the boys so they didn't mind doing it for me (but I had to rousie for myself) – I remember her telling me, "You better do the wool." And then I remember at this one place after being all kind of cranky at me earlier on, she was coming in on her lunchbreaks and rousieing for me . . . so that was pretty cool. I remember when she told me that I did a good job on one actually . . .'

The first school 'just starts you off', so Catherine progressed through the other training levels. At the advanced course with Phil Wedd and Mark Herlihy she asked to learn more about show shearing, which has held her in good stead as she competes. But that first course got her truly hooked:

'I did my first hundred at Dad's a few days later, I think. He set it up not long after that shearing school . . . and after the hundred I was buzzing. I'd met Bart [Hadfield] and he said I could have a stand at his place when they shore, and that was the next day after I'd done my first hundred. So I didn't get a break and my body was just killing. I did my first hundred . . . on these long-tailed stragglers that had been out in the forestry or something.'

Full of burrs, her arms and hands bore the brunt of them. 'Dad came down to the shed and put . . . compression bandages all up my arms; I think he brought gloves down, too, like I was red raw.'

She laughs as she recalls having 'a couple of little tanties' – but alone in the shed, flicking her own wool out of the way, Catherine remembers it not being 'the greatest day'.

Catherine recognises the important role that Bart Hadfield has played in helping establish her as a shearer. Introduced to him by a friend, she did a main shear up at Bart and Nuku's station, Mangaroa, near Wairoa. She remembers that Bart had taken on another couple of young guys, Adam Morton and Jason Hiko. Catherine thinks it was there that she did her eight-hour 100.

'They were on his ewes; I thought they were all boilers back then . . . huge and physical. I was buggered that day, really tapped out.

'Bart doesn't just shear when he's there; he's pressing, and he'll be out in the yards and he'll still shear heaps of sheep in the run – but he's everywhere doing everything. Me and Jas were shearing next to each other, and I remember he took us outside and he said, "Do you see around the ears . . . you're cutting ears." Which is quite easy to do, but easy to fix, so he fixed that. If you nick an ear it's not that big a deal, but they bleed all over everything.

'There's a way Bart says something to me . . . it just clicks. You can have a really good shearer and they don't have the right way of putting things. But I just pick up things when he tells me.'

Catherine reflects on what specifically it was that Bart taught her:

'Blows, definitely . . . the angle, for show shearing especially . . . Tectra teaches you a rough pattern, but he tweaked it to his own sort of style. I'd like to think that I still shear a bit like Bart; he's pretty smooth.'

Catherine passed her next major milestones at Mangaroa as well.

'Until I got my two hundred I wasn't working for anyone in particular; I was getting bits and pieces of work just to get experience . . . and shearing Dad's sheep probably over and over and over. Just trying to keep my hand in. But Bart got the tallies up. Yeah, I did my two hundred and I think I did my three hundred in that same stint that we were there.'

It was in the summer when shearing lambs that Catherine got her 200.

'And then the next day, Bart had taught us a pretty good lamb pattern, so I was just copying everything – well, trying my best to copy exactly what he had taught me – and then I think I did seventy-four or seventy-five that first run . . . and he said, "Three hundy day." I don't think I would have tried to do it if he hadn't said that. I remember him telling me – cos you don't know how to work to the clock or anything back then; it's all about the day, you're picking things up all the time – he was saying to me, you want to do this many every five minutes or by the quarter; you want to try for this many . . . It's really helpful to race the clock and have it broken down like that – it keeps you thinking as well, it keeps you focused during the run, otherwise you can float off and think about all sorts.'

Catherine is not keen to talk too much about tallies, but she knows that she can do a 'decent enough' tally now, having done her 400 lambs a few times.

'It gets easier and easier every year; you just learn. It's amazing when you look back and think, "Shit, I was doing it hard last year, for no reason." You just change something little and then it just comes easier. It makes more sense.'

Now that she is an established shearer, Catherine has created an annual working pattern. She works from the end of November through to the end of May for contractor Mark Barrowcliffe out of Piopio in the Waitomo District, then she is in Scotland for about eleven weeks. After that, she heads to Australia until it is time to return to Piopio. She is not afraid to compete, and there have been a few podium finishes – a highlight being the Lochearnhead win, especially as it was there that she nailed her nemesis, the Blackface.

'I really liked those sheep. I have a lot of respect for those Scottish Blackfaces. Ever since I went to Scotland I kinda wanted to nail them, cos at the start they can be your worst nightmare. They've got horns, and I had bruises everywhere when I first went there; a lot of people do. They can annihilate you, and they're not big sheep – they're just proud. If you're not doing something right they'll take advantage of you and destroy you. So, getting around them a bit better and getting the hang of them was a goal.'

When asked what she loves about working in the sheds Catherine is quick to reply:

'The craic – I don't mean the drug crack, just the craic. There's the funniest things . . . your sense of humour changes and the random things will be funny; I think there's a shed sense of humour. Like, before Christmas, before we knocked off we were at the shed, and this guy – he's a hard-working guy but just full tilt all the time, and kind of uncoordinated – he came in, and we must have been having a wool change off the lambs and onto the ewes or something, so he was helping with that and trying to help the presser, and he was up in the press tramping. He must have jumped out, but he tripped and fell onto a pile of wool. I heard a big thump and looked up . . . and I saw that he had fallen on the wool, so I wasn't worried cos I knew he was alright, so I carried on, and then you just hear this "man down", and I just couldn't breathe . . . you always have something funny that happens, you know . . .'

There is talk of her trips overseas, working in Australia and shearing in the Barrowcliffe gangs. As Catherine talks there is little trace of that shy wee girl at the woolshed; rather, there is now a well-travelled and competent shearer who still sees room to grow:

'I don't think I'm at my peak; I wouldn't say that. It's definitely a lot easier on my body and I'm a lot happier in my work, cos it's not hurting so much.'

If Catherine is at the beginning of her journey, shed-shearing legend Eddie Parkinson is getting quite long in the tooth. With over four decades in the industry, he is a veteran. With his competitive nature and drive, and ability to churn out the numbers cleanly, he earns the respect of all those who work with him.

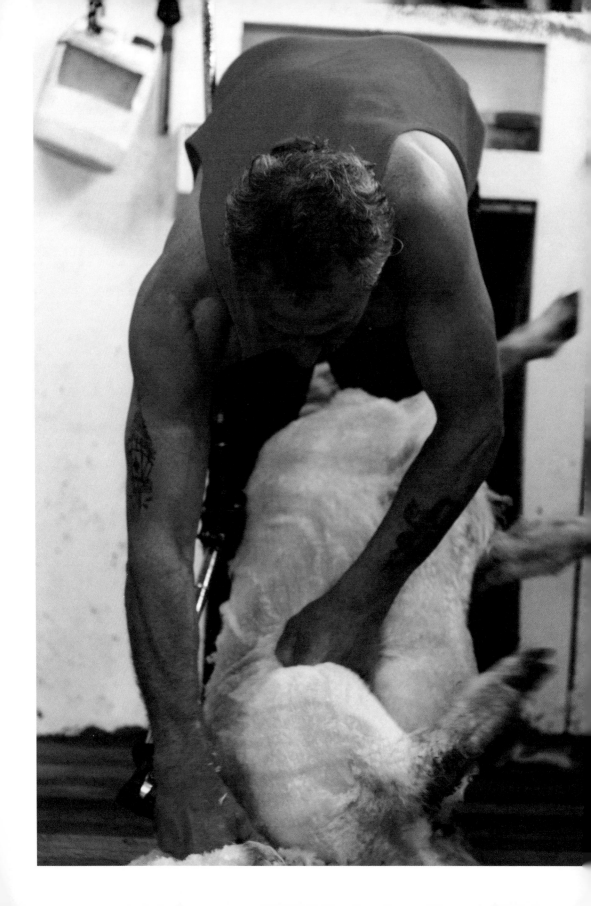

Eddie Parkinson –
A shed-shearing legend

*'Everyone who works in Taumarunui, all they ever
come back talking about is Eddie Parkinson this and
Eddie Parkinson that – he's a legend. Even when I was
at Lochinver, he was down the back shed, he had to
drag the sheep about ten feet across in that back shed ...
There was heaps of gun shearers there and Eddie
shore the most sheep of anyone at Lochinver that year,
and he probably had the longest drag. He was quite an
incredible man. He's the smallest man stature-wise by
a long shot, but mentally and physically he's the man.'*

E ddie Parkinson – or Parky, as he is commonly known – has been
shearing sheep for over forty years, twenty-seven or so of them with
Mackintosh Shearing. As Bart Hadfield comments above, when it comes
to shearing, he's the man. Parky has shorn close to two million sheep, and
at sixty-three he is still going. He may like the occasional day off now, for
game fishing and hunting, but the handpiece is not being hung up just yet.

When Parky talks about shearing he speaks in terms of hard

work, drive, competition, effort, focus and goals. His numbers speak for themselves, and he has been a fast and consistent shed shearer right from his earliest days on the board aged twenty-one. He credits his driven and competitive nature to his father's influence.

'My dad, he worked and worked hard; that was instilled into me at a young age, and to be competitive no matter what you do. He never wanted to come second or third. Even when we were pig hunting you wanted to be the first one up the hill with the pig – it drives you.'

Brought up on Robinsons' station at Ōhura in the King Country, where his father worked, his early life was 'pretty basic' but a 'good life': roaming around on horses, pig hunting up the Waitōtara Valley with his father, milking cows, chopping firewood and feeding the chooks were all part of the mix. Shearing was, of course, part of the station calendar, so Parky was introduced to a shearing shed early. 'I used to sit up on the bales; I was only a little rug-rat sort of thing, but I used to sit up and watch them.'

After a family shift to Bulls, Parky took up school holiday work, penning up and crutching with Palmerston North contractor Colin Trembath, for whom his mother had rousied. After leaving school aged fifteen, before heading into the sheds, Parky tried his hand at post cutting at Santoft Forest, and was the top driver at potato grower Jim Bull's for several years. He had even flirted with the idea of becoming a greenkeeper, but when he heard the pay was just $14 a week he decided piecework at Santoft would earn him more. When his job with Jim Bull changed, he decided it was time to try shearing.

Colin Trembath was the first to give Eddie a stand. 'I must have been a good worker, he must have seen the potential – I'm not putting myself up there, by no means, but he just gave me a go and in a few weeks I'd shorn my two hundred.' Parky remembers Colin giving him a bit of gear, showing him a bit, and then he was left to get on with it. It was 'dog eat dog', with older shearers not keen to pass on information initially. 'You're taking their work away in those days, cos there was a lot of shearers around, a lot of freezing workers, and fellas used to do the season there and then go shearing.'

Parky's first day shearing was memorably tough, but he got stuck in:

'You've just got to concentrate on what you're doing, focus on what you're up to, try and make some money; if you didn't work there's no money. I just about died; it was tough, oh it was sheer hell, but it taught you a lot of things.'

Parky laughs at the memory of his first sheep: they were 'quite rough', but he adds that quality comes with time. 'You had to have a certain standard, otherwise you'd get the sack.' Parky's milestones came quickly: his first 100 on his first day, his first 200 a few weeks later.

'Oh, it's quite an achievement, those first one, two, three hundred – the hardest ones you'll ever get cos you can't really shear a lot; it's more physical, less technique . . . You have to watch and learn, and the faster you can do that, you can achieve your goals . . . I always had visions of shearing heaps of sheep – it was just a stepping stone for me.

'Three hundred's quite a milestone for a shearer. It's probably the hardest one because you're in that period where you still don't know how to shear properly. You haven't got that experience . . . there's still a lot to learn. People think just cos you've done three hundred you're the man, but you're way off the mark.'

Parky is matter-of-fact about reaching his 400 and 500:

'That just comes a lot easier, four and five hundreds – quite a lot easier, I think. It depends on what you put into it: some people never get there, they never shear five hundred . . . I was fortunate I got there after a couple of years of shearing. I don't know – you want the will to try and be good at what you're doing. If you shear sixty in the first run – say, for instance, I want to shear sixty all day, I don't want to drop down to forty or fifty – that makes your body that much harder, and then you learn the mental things. You'll get a bad pen and you'll end up two or three down, and then you've got to dig in and get back up to that number, whatever that number may be . . . I don't like to say it's in the too-hard basket; that's just the way I am.'

Shearers are very good at acknowledging those whom they admired and learned from. For Parky it was Lofty Crawford:

'He was a very talented shearer. I always watched him going round the sheep, and I eventually ended up beside him shearing. You watch places on the sheep, you try and beat him; if you can beat him there, sooner or later you'll beat him. You're studying people and picking up their knowledge.

'Once you get on to a mob of sheep, in the first, say, quarter of an hour you should know how you are going to shear those, within reason. You can lessen your blows because they're smaller sheep, or . . . you try and shear to a pattern and get round them as quick as you can. You don't shear every sheep exactly the same. The bigger sheep, you might have to do five blows on the belly cos they're full and a lot wider and the trick is to try and make your sheep smaller . . . by footwork and various other things you can make your sheep smaller with less blows.'

Parky's opportunity to start really churning out the numbers was when Bill Morrison, a big Wairarapa contractor, got wind of the young shearer and approached him.

'Like, I remember when Bill Morrison came and saw me at the Golden Shears he guaranteed me 20,000 sheep before Christmas. And he said, if you can do 20,000 sheep I'll give you X amount of dollars more. And that happened. He was a good operator . . . it's a numbers game: the better shearers, they make him a lot of money, so it's in their own interests to get good staff. And that's how it works.'

Parky had twelve years with Bill averaging 400 a day. Alongside that he worked down with Elbert de Koning Shearing at Wairaki and Mount Linton in Southland; with over 60,000 ewes on offer at Mount Linton there was plenty of work. He had an early stint shearing merinos at Kurow in the Waitaki Valley, as well as over a decade of travelling back and forth to Australia working for two- or three-month blocks – at one stage he worked eighty days straight. To him it was all about 'good shed shearing consistently'.

Parky remembers there being only two brands of handpiece when he started: Sunbeam and Lister. He turned to the older experienced shearers for advice and picked their brains.

'For a start I had Listers, and then I went to Sunbeam EBs. Sunbeam seemed to be better handpieces, they were smoother, they lasted longer.

'Gear's come a long way. In the seventies and eighties you had to do a lot to your gear to get it running, and that gives you the edge – thinning them out and spending time on your gear. It was really important. Thinning the tips out so it enters better. If you knew a lot about gear it was to your advantage. When I came up here shearing I was fortunate enough to shear with a guy called Robbie Tai, quite a well-known shearer, and we used to sit down after work and muck around with gear. He taught me a lot, and that enabled you to shear a lot more sheep. You'd put them on a spinner and then spin them down and reshape your tooth just to make them go in better, and you have all different combs for different types of sheep. I've probably got eighty combs and they're all different: there's new ones, there's half-worn ones, there's thin ones, and you need all of them for different types of sheep, so when you look at the sheep you can put something on that will go. If they're bony, for instance, and skinny, and you can see that they're tight, you need something thinner to get in there and try and get it off. And if they're nice and round – well, you can put a new comb on them and get round them a lot quicker. There's quite a lot in it, but that comes with experience. Combs have come a long way, and you don't have to do gear up to the extent we used to in those days. A lot more accessible: just get them out of the packet.'

Parky kept peeling wool off sheep at an impressive pace and has never been short of work.

'I think I shore my first six hundred in my third year shearing. I got up there really quick. Six hundred lambs – and I shore five hundred ewes that year, too.'

His reputation as a gun was truly established.

'Fagans [brothers John and David] were here at Taumarunui and I've got a good friend, Neil St George – he was working for them and he was breaking a few records round here. I was working for a different fella and then people heard that I'd done 620 . . . So people want you to shear for them . . . but then you're chasing numbers. You're there to make money: if you can shear for good contractors, it works both ways. I was just fortunate and a bit gifted in what I was doing, I presume.'

When Ewen Mackintosh, founder of Mackintosh Shearing, started out in 1990, he approached Parky to come on as a permanent. Parky remembers weighing it up. 'Are you going to be better off? You've got to get those numbers of sheep – the more numbers, the more money. It's no use going to work for someone just because he's starting the run up.' Ewen had to ask a couple of times before Parky was ready to make the move, but it is safe to assume, considering Parky's long working relationship at Mackintosh, that he has had plenty of sheep to shear.

While numbers have always been important to Parky, it hasn't been at the cost of everyone around him.

'You're all doing your own thing: you're trying to achieve as many sheep as you can get out the porthole yourself. But you still look after the people you're working with and be respectful to them and look up to them. You got to look after the cook. Everyone's got to gel, otherwise you can have very unstable times.'

Parky explains what has driven him all these years:

'You don't want people beating you – well I don't, and I still don't: they do beat me now, but I still make them work hard for it! It's just the way you're made – it's just something in ya – I don't know what it is.'

He also voices the frustration in the attitude of the young shearers today:

'The shearing industry these days is nowhere near as competitive as it was twenty years ago. I think people just treat it as a job – a lot of them don't respect things. When I was learning to shear, on an eight-stand shed, the slowest was on number eight and your goal was to get to number one as quick as you could. But that doesn't happen these days. They don't want to achieve – they don't want to get up there and be number one. Some do, but the percentage is way down. It was like a racetrack when I started shearing, it was very competitive. And if you could beat a fella three days in a row he went back on your stand and you moved up, and you stayed there and then you moved up again. It was that respect thing – whereas that's all gone out the door these days. People walk into a shed and think they can jump on number one – well you'd never do that, you'd get a clip round the ear. It just wouldn't happen, out of respect, you know: the best shearers, they earned that. That has drifted out of the shearing industry to a large extent.'

Parky reflects back over his years shearing and counts himself lucky to have had the input of two really gifted shearers:

'That guy Lofty Crawford was a very talented shearer, and I was just lucky to be there when he was in Palmerston. He was very fluid – he was one of those Māori shearers that were just gifted, you know . . . he was one of those guys. And Robbie Tai: gifted in doing gear, and I just happened to be in the right place at the right time and picked up that knowledge. You've got to want it.'

And Parky laughs. 'There's no easy way getting round a sheep.'

A theme that comes through loud and clear when talking with shearers is that shearing is as much a mental game as it is a physical one. Mental toughness, which Parky oozes, gets a shearer through the hard days or when spirits are sagging as they drag yet another sheep out of the pen. They also need mental agility. When the handpiece is not 'cutting like a hot knife through butter', they need to work out how to adjust their gear to suit. There is also a constant adjusting of their shearing pattern to suit each sheep. How you tackle a Perendale is going to be very different to how you tackle a merino. Barry Pullin talks through some of the mental gymnastics.

Eddie Parkinson is always willing to impart his knowledge to those who show they are keen and are willing to work hard.

Barry Pullin –
The 'top two inches'

*'To get the best out of your shearing you've really
got to sit down and analyse how you shear the
sheep and what you're doing and where you're
putting the sheep and where you're putting yourself.
So, physically it's hard, but mentally it's very hard,
too, and you don't realise how much energy you're
burning because of that mental concentration.'*

'**P**iss-poor planning promotes poor performance' – or the six Ps, as
Barry calls them – is the mnemonic by which he lives. A maxim he
applies to all that he does, it reveals his analytical nature and his emphasis
on engaging the brain. To date, Barry has been a shearer for thirty-six
years and a Canterbury shearing contractor for thirty-three of those
years. He and his wife Trish took over a run from Allan Ford and have
developed it into a very large and successful business. Barry studied for

a BCom Ag from Lincoln College after leaving school, and although the commitments of a mortgage coupled with the desire to get on and shear derailed him from completing it, it gave him a foundation. He was also a successful participant in the Kellogg Rural Leadership Programme in 2006. The paper he produced for the course, 'The New Zealand Shearing Industry: A case study', gave him an opportunity to analyse the labour force characteristics of the industry using his own business as a basis for his research. The principles he distilled through the paper demonstrate his analytical bent and have guided his management practices ever since. For Barry, shearing is as much about using the 'top two inches' as it is about brawn.

Above the garage of Barry and Trish's Rolleston home, the base for Pullin Shearing, is a well-appointed man-cave where Barry holds regular meetings and workshops with his crews. The air-conditioned room is a comfortable space to talk, considering Canterbury is turning on the summer heat at thirty-plus degrees. Barry grew up on the family sheep and beef farm in Purua, Northland, where his mother Lynette taught him the rudiments of shearing. Barry's father suffered from ill health, so Lynette, a trained nurse, took up the farm work, doing some farming courses at Massey and teaching herself to crutch and shear.

'She was a very strong, very determined, very stubborn pain in the butt at times. She used to chase us up the tree with a bit of supplejack.

'Mum shore with us, or she stood there and pointed out with a bit of stick and tapped you on the hand and whatever or gave your leg a belt and said move it here, move it there . . . When I started to shear away from the farm in gangs or in crews, you learned off the other shearers: you'd watch and see what they would do.

'Mum was very good at controlling the sheep. We have employed quite a few female shearers over the years, and one thing they are very, very good at – because they have to be – is working with the animal and having it in the right place. Instead of physically dominating the sheep and putting it in place for you to shear, you work *with* the animal. You understand that if you press a point on the front shoulder the leg will straighten out; and if you put your finger in down on the hind leg, just behind the knuckle there, the leg will go out . . .'

While Barry is willing to admit there are times where brute strength might be necessary, it's not an approach that is sustainable.

Most shearers can remember those early experiences of holding a handpiece, and Barry is no exception. He notes how the antiquated handpiece and combs were very different from the modern alternatives:

'I would have been twelve or thirteen . . . you had this thing that was going to chatter away in your hand and it got extremely hot, and I can remember having handfuls of blisters – mainly because I was probably cutting more air than wool. And then you had this sheep that you were trying to hang onto at the same time. I couldn't get it together properly for quite a while, till I was probably thirteen or fourteen. At that stage I was learning to shear – crutching and shearing. The handpiece was an old EB Stewart, a big fat horrible thing, one of the original Sunbeam handpieces. And there were some Lister handpieces – they were like the modern handpiece but a lot wider in the barrel, and often had a rug around them. They would have a leather glove or leather stitching wrapped around the handpiece so that when the handpiece got hot it didn't burn your hand. That basically came from Aussie, and they were called a rug. We Kiwis cut these leather wraps or rugs off and we hung onto the steel. The leather would get extremely slippery when you were sweating so much, and it was easier to hang onto the steel. You'd burn your hand if you didn't keep it in the wool when it got hot, but when you are a young fellow it is going to cut the wool off no matter what.

'Combs and cutters were not as good as they are now . . . basically we had some Lister combs and Lister Tally Bs – they were a big wide comb – and then there was only three or four Sunbeam combs, then Final Hustler and Top Flight and that was about it. Now each of my shearers has probably got about fifteen different types of combs. We were shearing more consistent sheep, breed-wise, so we didn't need the range of combs that our guys now have to have here in Canterbury. And we learned how to get the best out of a comb and adjust it, trim it and shape it to do what we needed it to do.

'When preparing your combs, you're aiming to – the bevel is the roundness and the point that you have got on your comb, so if it is earlier or stickier shearing you want something that is a little bit pointier; if it is a fine wool you want something that is a little bit pointier. The scallop is the wee bit that sits on top of the comb . . . it allows the comb to slice in and then the scallop parts it before the cutter hits it. You would trim and adjust the scallop to come back in a comb so it hasn't got a sharp edge. (If you pick up the skin it will cut through.) Timewise, to really get a comb going properly you could probably spend five to six hours on a comb with a bit of paper and sandpaper, fine sandpaper or a wee spinning disc on a drill.

'When I first started shearing – and a lot of our shearers are that age – we learned how to adjust the comb by hand ourselves, and it certainly paid dividends when I was starting to shear pretty hard twenty-five years ago. The modern shearer today buys a comb straight out of the packet and they do very, very little to the combs.'

Before he started contracting, Barry used to find grinding the gear at the end of the day quite therapeutic – and perhaps for the analyst in him it was an opportunity for further evaluation.

A mate taught him to drop his used combs and cutters onto a needle in his gear bucket as he finished with them. It kept them in order of use.

'I quickly got into the habit mainly because I could actually go back and look at the hours that I'd shorn, and think back and say, okay, on the second run those two cutters didn't work so well, or, that was a bad comb choice. So I could go back and see what was going wrong – whether I had bevel or the scallop wasn't right, to see if I ground it properly.'

The analysis does not stop with the gear. Barry explains the dynamic a shearer engages to increase their tallies:

'I got my first hundred within my first week. It felt pretty good, and I couldn't wait to do my first two hundred. You go in stages: when you start off, your focus is very, very wide. When you are aiming to shear a hundred – now this is going to sound stupid, but – you look at shearing your first sheep. To do your first hundred your focus is so wide, and it is such a big task that you look at shearing five runs of twenty. So you are not worrying how long it is going to take to shear a sheep, but, as long as I get twenty sheep done in that first one-hour to two-hour run – or twenty-two. Or, if it is an eight-hour day, I need to do twenty-five. When you shear your first five hundred you think, how long is it going to take me to shear a sheep? And I know that I have to be whacking it out in about fifty-six seconds . . . so it is an inverse relationship to how you plan and how you look and your mental approach. So . . . when you want to go to truly macro at five hundred you have got to go micro to do it. To really nail your five hundred you have to have the individual part of the sheep and the pattern of the sheep maximised – so you have to look where every blow is, every foot placement . . . the things that add up for you to shear a sheep consistently in fifty-six seconds from catching and dragging it out become so crucial. It's an inverse relationship – so that's why when you start getting better and better and better at your shearing, you forget how many you are doing and you look at the individual sheep and where it is, and if I can move my foot, my toe weight from the outside of my little toe to on my ball and then move my heel round a bit I can get a blow in a bit easier . . .

'Some guys will never, ever get it because they never, ever understand that to shear more they have actually got to look narrower. They look at 250, or they will look at sixty a run on an eight-hour day and say, "How can you shear more? You just can't physically do it," and I say, "Pretty easy to do when you start using your brain." It's the same as any top sportsperson. To make the All Blacks the way they are, they look at the micro and at what builds every single facet of the game to accomplish the bigger goal. It is exactly the same as the marathon runner; it is exactly the same as the sprinter, the cricketer, whatever . . . Once you develop that awareness, you develop your consistency, so you become very strong mentally – so that your target will go out no matter what.'

Barry also points out that when you're consistently shearing 200 sheep a day, patterns click in:

'You start to understand that you are looking at your counter or knowing how many sheep you have done in your time, so you are starting to pace yourself against a watch. To shear two hundred on an eight-hour day is fifty sheep every run, so you want to be shearing twenty-five in the hour . . . or a sheep every two minutes fifteen. Every five minutes you want to be doing two sheep and one up the neck probably, so you're starting to understand time management. What's going on is that you are starting to really focus and understand where that pattern is and where that blow is and what position you are putting yourself in and what position the sheep is in, so you can make the best out of it.'

Barry acknowledges Wool Board instructor Peter Burnett for helping him get to a point where things clicked:

'And you suddenly go, "I am a brain-dead idiot! This is not actually that hard." It becomes a bit easier, but it has come a lot easier because of your approach to it. It's really when you start thinking and start using your brain instead of your brawn that you will start going up in your numbers . . .

'I always break the sheep down into sections, and I could go the same as a

LEFT Boss Barry Pullin keeps his hand in and shears one of the Muzzle Station wethers, giving a young shearer a chance to stretch his back and garner the finer points of shearing.

TOP RIGHT Head woolhandler Lashara Anderson gets ready to throw a fleece. It takes skill to throw a fleece well, especially when they're almost as big as you are.

BOTTOM Team leader or ganger Peter Hay works his gear on the grinder to ensure it is in optimum condition for the next run.

beat on a song or something. I would say, I want to do one-two-three-four here – where can I get to? And I would go one-two-three-four plus one . . . and you just develop. It is part of your rhythm: some people get it naturally, some people have to really think about it, and I always broke it down into sections. As I got better and better I would start counting my blows on the sheep and then saying, okay I can get round this in forty-six to forty-eight blows . . . and then I would watch somebody and think, okay, he is doing it that way – I will try that.'

As his style evolved, Barry became more efficient.

'Blows usually get less – so, instead of putting in a lot of short blows and a lot of half-blows, you will start putting in bigger blows and maximising them. You look at getting efficient in your pattern. It means you are not cutting the same area twice, either by going over a shorn bit of skin with a comb with no wool or cutting the wool off twice, so that is a second cut. Get rid of your second cuts and fill up the comb, and you are suddenly getting a lot more efficient.

　'People think a lot of shearing is just sheer physical hard work. No, it's not. An average full-on concentration for twenty minutes is bloody hard; when you are doing it for two hours it is very hard, but sometimes to get better you actually don't think: you go fishing – you think about something else. When I am teaching and showing people here I say, "Try do it, forget it, and then come back and try it in half an hour or at the end of the hour. And if you like it and it is easy for you, you'll naturally start doing it." It's patterning and sequencing, and you have actually got to relax and pull back from it to allow for that sequencing to cement in. Visualisation is a really good key; if you are learning to shear you should be visualising how you are going to shear that sheep before you shear it – not each sheep, but definitely before you start work in the morning or on smoko. Or you should be running through it and thinking, "If I want to get better here, how am I going to do it, how do I visualise it, how do I preplan it, where am I going?"'

There are other factors where mental adjustments have to be made. Regional variations in sheep can have a huge impact on how a shearer performs, and Barry had to adjust to Canterbury sheep after shearing the Northland sheep. Barry explains that in Northland, they were 'a cross-bred up there and they are quite open in the points with a lot less wool on them. Down here we are dealing with a lot more wool per area; more Corriedales, more of a finer wool – different gear needed, a different approach . . .'

'When you move from a second-shear cross-bred ewe in the North Island and you come down into the northern half of the South Island or right down through into Kurow or places like that, you're going from a cross-bred situation to more of a Corriedale, half-bred fine wool. So while your tally up north might be three hundred, you'll be scratching to sit on two hundred until you get your head around the sheep and get your gear sorted – they're just so much harder to shear. As you go further south and you go into the bigger, more open-wool cross-breds that they shear in Southland and Otago, your tallies will go up because they're easier – they're more Romneys, Romney Marshes, and they're cross-bred – there isn't as much work in them; you're not pushing off the same dense fibre. Canterbury's a pretty unique place: it's a great place to learn and it's a great place to come back to because of the continuity of work, but if you're going to shear big tallies it's not the best place in the world to be because it's tough . . . yip, it's hard. If you go from here up north, you'll be looking at plus one-fifty, plus two hundred; so if I shore three hundred in Canterbury, the odd time I'd go up north I'd be looking at four-fifty, five hundred, five-fifty. If I was going south, three hundred here was an easy four hundred down there.'

Barry takes time to explain just what he means by the density:

'What you are looking at is micron. A micron is one-thousandth of a millimetre. Coarse micron is cross-bred, so you are looking at anything over thirty-one, thirty-two microns. Going up becomes coarse; the coarser the wool, the harder-wearing the fabric. So woollen carpets are made out of thirty-six, thirty-eight micron; you'll get some finer stuff in it to give it the soft feel. The coarser the wool is, you get what we call the prickle factor, which is the itchiness or the bite that you get when you put on a homespun Romney jersey: unless you have a skivvy or something on underneath, it is going to bite you on the skin. Once you start going below thirty-one micron down to about twenty-seven, you are on the fine edge of your cross-bred, your Perendale, and into the softer types. Technology in wool processing gets rid of the prickle factor, especially on that finer edge, so you can get some finer babies' blankets, some finer apparel out of fine cross-bred. Twenty-seven micron and below you are getting into what we call mid-micron, which is your Corriedale. At twenty-three you are starting to get more into apparel wear, which is the good half-breds and quarter-breds. And once you go below about twenty-one, you are in the merino. The finer the merino, the closer it is to your skin, so below twenty-one you get into the leisurewear and into smart wools, Icebreaker . . . and from eighteen . . . to twelve you get into very fine suiting wear. We have just done one superfine clip here on hoggets; it is 12.2, which is very fine, like silk. You are looking at around $42 to $44 a kilo.'

It took Barry a while to get his head around the change when he settled in Canterbury.

'You pick up the rudiments pretty quickly, but it is probably three to four years (or seasons) before you start thinking, why does that comb go on that half-bred? And what makes it suitable for that sheep? And it wasn't until I ended up having my own business and had some really good shearers that we started focusing, and I really understood the importance of gear.'

The mental energy that Barry put into his shearing is now very much directed to managing the running of Pullin Contracting. He reflects on whether everyone sees shearing as such a mental game:

'Some people will understand the concept and start using the "top two inches", so to speak. Others will just bash the wool off whichever way they can – do it physically, dominate the sheep, try and dominate the area around them; but those people hit their ceiling pretty quick. Those who want to extend themselves and keep looking for their ceiling pretty quickly understand that you've got to start focusing in.'

When you listen to Ian and Beth Kirkpatrick's story it becomes clear that they have spent plenty of mental energy 'focusing in' – not just on the shearing itself but also on making the most of the opportunities that shearing brings as they've set up their contracting business. Along their journey they have been giving young lads the opportunity to take up the handpiece. Ian invests time in them, training them in the shed, enabling them to reap the benefits of their hard work.

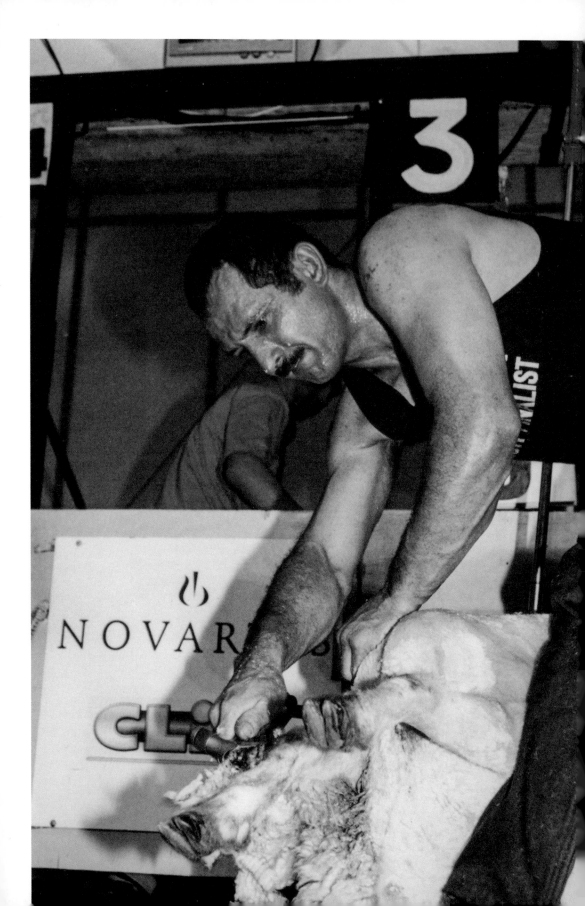

Ian and Beth Kirkpatrick – 'Making shearers for life'

*'I feel we were lucky that our dad taught us to work –
and you just tackled whatever was in front of you.'*

Ian and Beth, busy Gisborne shearing contractors, have generously stolen some time out to be interrogated, turning cellphones off so they are free from distractions, and putting the sixteen shearers and shed hands currently working for them (as well as their clients) on hold. Not knowing quite what to expect, the couple understandably look a little apprehensive entering the lounge of client Geoff Candy. They are probably more familiar with the inside of Geoff's woolshed on his Marika

Station at Rere. But, being obliging, they sit ready to answer questions about their lives in the shearing industry.

With the surname Kirkpatrick, the first question is obvious: Are you related? Yes: Johnny, the current world champion machine shearer, is one of Ian's younger brothers. Ian is proud of his brother's achievements, knowing that he has worked hard.

'All those gun shearers – my brother Johnny, Rowland [Smith], David Fagan – they strive to be at the top of that ladder, and what you put into it is what you are going to get out of it.'

Beth and Ian are in their early fifties and have been contractors since 2001. They met at high school at Tolaga Bay and have been together ever since. They started a family young and Beth comments, 'We had to grow up fast.' It was definitely a no-frills life early on, and they weren't afraid to work hard to get ahead. Ian says: 'I look at our life, and me and my wife – we left home at fifteen and we lived in a little bach. We had no power, no bath, we cooked on an open fire. We had one room and there was a spring across the river; we carted water from there . . .'

Beth and Ian laugh together at the memory of one particularly 'hard-case' day:

'Back up the coast I shore and got on well with a lot of farmers, and we lived in Tokomaru Bay. There was this shed right at the back, just about in behind Mount Hikurangi, and the farmer had some sheep to shear. I don't know how old our little boy was [Beth adds that he was just walking]. I had a motorbike, a Honda XR trail bike, and so we went up there for a five o'clock start. Me and my wife and our young fella jumped on the motorbike – he was tucked in the middle – and we rode off to work. And it was just me and my wife at the shed all day, then jump on the motorbike and come home . . . our young fella cried most of the day, so that sort of made the day quite daunting.'

Beth adds, 'It's a day I won't forget in a hurry.' She was fleecing, but remembers it wasn't easy with a needy child. 'It was quite hard because you had the little one there as well, and that was a long day.' Ian: 'And you know, the road's quite windy once you get up into those back blocks.' Beth: 'We were sliding here and there as well – that was crazy – I can't believe we did that.' Ian: 'But it was cheap travelling; you jump on the bike and away we went.' Beth: 'And it was freezing.'

Ian begins to talk about his childhood, and you quickly learn that a good work ethic has been instilled in the Kirkpatrick boys. He quickly learned the benefit of hard work, earning money cutting thistles and scrub when young, and even remembers shearing his first sheep at six or seven years old, when the family lived up the Waimata Valley on a farm his father managed.

With a move to a family farm in Te Karaka, Ian learned to shear:

'My dad and his brother were buying the farm and we done a lot of the work ourselves. I sorta shore my first two hundred when I was twelve . . . working on a farm you learn to work hard and . . . at fourteen or fifteen I shore my first three hundred, at sixteen I done four hundred – yeah, just worked away.'

There is no proud puffing of chest as Ian talks; he almost brushes the tallies aside and states that there are a lot who could shear 200 by age twelve. He explains how he came to achieve at such a young age:

'A lot of that is just determination. You know, you've got brothers who could shear – I had an older brother, he was a shearer and he went away – he left home at fifteen and was away shearing. It's like life's a competition, and I just feel we were lucky that our dad taught us to work . . . you just tackled whatever was in front of you. You shear with fellas, and if they're quick you want to be as quick as them: you've gotta work to be as quick as them, that's what I put it down to. We didn't learn a lot about shearing, but because we could work, that helped us.'

He admits there was no finesse about it: he would 'hook in and just grunt away'. Ian acknowledges the help of Chris Teutenberg, the instructor at the Te Karaka and Tokomaru Bay shearing schools the Kirkpatrick boys attended, in refining their shearing technique. Ian even remembers Chris spotting Johnny's potential right back then.

The work ethic instilled in Ian at home carried on into his working life. Before going shearing full-time, he shore for the main shear, then worked in the fields:

'. . . in a squash crew picking pumpkins January/February, then slipped back into

second shear. And when the forestry boom was full-on around here I went planting
pine trees. That was a good job – I really enjoyed planting pine trees – out on the
hills. And if the weather was wet you could still go to work, you could put hours
in – whereas with shearing, once it's rained you had to wait for the sheep to dry.
I planted a lot of trees and made a lot of money out of that as well, then as the
journey went on we started shearing a bit more.'

As Ian talks, you become aware of just how driven he is. For one thing, he
hates taking days off: 'It kills me.' Beth: 'If he's not shearing he's always
doing something else around home.' Ian: 'I learned to weld and build and
renovate houses, and we've done a bit of that and made a little bit of money
out of that as well. Good life skills, I feel.' And as for taking a sickie, Ian
recalls a time when he injured himself shearing and went to the doctor's
to get stitched up and headed straight back to work. He jokes, 'I started
thinking I wish the doctor could give me a needle and a thread and I'd just
stitch myself up at work and carry on . . .'

When asked what drives him, Ian explains how he set his sights high:

'Some gun shearers have nothing at the end of their shearing career, and I feel
when the body's had enough you want to be able to walk away and have something
of your own. It's no good having nothing and then just becoming reliant on welfare;
I feel you've got to set things up so if the body gives up, you've got things that will
see you through the rest of your journey through life. I feel if you're there to work,
you go along and you try and stay focused on what you're trying to do: don't get
tied up in too much of the hype that everyone else is doing. It's a tough one for a
young person: they talk about peer pressure, and we slid down the drinking road
with all these other shearers; but you look at it and think, nah, that's not me: my
sights are set a little bit different. And so you just go down the road you want to,
and if they think you're a little bit strange you just keep on going.'

After twenty years in the industry, contracting was a natural progression
for Ian and Beth. Ian's built up a reputation as a good shearer and has a
record of working hard throughout the country: Mount Linton, Ngamatea,
Otupae and Ōhai were all part of his regular north–south pilgrimage.
There were Landcorp farms down at Te Anau when he shore for George
Potae: 'I was down there for about two months and only shore at three
sheds. Big sheds, you know: there was fifteen thousand for three of us at

my first shed, and you park up there for twenty days.' There was a Western Australian adventure in there, too. Ian jokes that it was after the Wide Comb debacle: 'Kiwis had just about taken control by then.'

The contracting has grown, and while initially he had to call in whānau to help him out, now he has an established core of shearers and shed hands. At peak times they employ twenty to twenty-four shearers plus shed staff, and they'll have five or six vans out. Ian and Beth explain together:

'Just about all the shearers we've got are young boys that grew up with our young fella.' Beth: 'They didn't want to be at school.' Ian: 'All townies; we took them out and they learned how to press and sheepo, and then we taught them how to shear – and they've turned into good shearers. There's one young fella there, he used to come over home, and because our young fella was already working, he had all PlayStations and whatever he wanted to play with. And this boy, that's all he wanted – to come over and sit on the PlayStation. He wanted a job, so I gave him a job – my wife said to me, "What did you give him a job for? He's only good for playing on the PlayStation." But I feel you can mould people into something.' Beth: 'He's still with us today.' Ian: 'He's turned into a good shearer and he's a good young fella . . .'

The work can be stressful at times, as Ian explains: 'We work all day, and then at six o'clock or half past six you get home and you might be just having tea and then the phone starts ringing, and the phone usually rings through till about nine o'clock at night, and then it drops off. And then you get a little bit of family time, and before you know it it's ten o'clock and then you hit the hay. Usually I get up at four o'clock and go and grind up some gear, have breakfast, and five o'clock you head off to do your pick-up.' Beth: 'Or you're getting woken up by texts or phone calls – from workers.'

But the hard work doesn't put them off, and Ian relates another story that shows just how far he is willing to go to ensure the job is done:

'You've got to love it, eh. Sometimes we get under a bit of pressure with farmers, and this season, with the way it's gone . . . it's hard to get dry sheep, and then as soon as you get dry sheep everyone wants you, and we had a farmer up the back and he wanted nine hundred lambs ring-crutched. I spoke to one of my sons and we went up there and started about half past three in the morning; we were somewhere else shearing and I said to the farmer, "We'll be there three o'clock . . ."

And me and my young fella, we crutched for four hours straight through. We finished, I don't know, seven o'clock, and hit the road and started back at nine o'clock with the other crew. But you just do what you got to do.'

Ian is keen to invest time in his young gang.

'You teach them how to set their gear, and then teach them to shear. A lot of the younger guys, I like them to shear next to me . . . I can hear them having trouble and look around and see how they're holding their sheep, and then just go over there once I've finished, or watch them for a moment and say, "Righty-o, when you're having trouble back there if you just put your foot – cos sometimes it's just the way you hold your foot. If your foot is pointing out, you can turn it in and . . . push your knee forward and you can lock that sheep up a bit better. That's what I do, and I'm a great believer in it. When I started, the old guys showed you nothing, and I'll show these kids as much as I can cos when I leave this world I'll know that I've left behind the knowledge I had.

'I go over and watch them and see where they're roughing it up, and . . . help them tidy that spot up. I'd rather tell a shearer off than have the farmer come in and start growling them. You could say, "Righty-o, there it could be a little bit rough around the sheep's head," and you watch and see how come they're a little bit rough. When you can shear yourself you can say, "I'll shear a sheep and I'll show you" – you know how to clean that spot up and usually that's enough. I'm quite tough on all these young fellas: they've either gotta do the job properly or there's not a job there for them.'

Beth comments that the boys do shows as well. Ian explains the benefits: 'One thing we've encouraged them all to do is have a go in these shows; and if you're gonna do shows you've gotta learn to shear properly. And then they bring it back to the sheds, and all the farmers see these boys shearing clean and you've got no worries.'

The boys Ian and Beth take on are mostly Māori and perhaps haven't had much direction in life; the couple see that shearing gives them 'a bit of direction – they'll be shearers for life'. And creating shearers for life works within Ian's long-term vision for his contracting. For the Kirkpatricks it's about giving opportunity: 'Most of them are hungry to learn; I feel some will make it and some won't, but as long as you give them that opportunity they'll decide for themselves whether that's what they want to be doing.'

Ian is not always shearing these days. He may come in and do the pressing, and it has its frustrations.

'I stand back and watch the workers and they're just not hungry enough, and it kills me. I think, I could be on there shearing a lot more sheep – but you've got to give them their chance.' Beth comments that when Ian works alongside the young ones they all try to beat him; 'Some of them are quicker than him now.' Ian laughs: 'Some of these young fellas, like, I might beat them one run, and if they're shearing close to me I'll say, "How many did you do, bro?" And they'll tell me, and I'll go, "Gotcha." It's all good fun: it keeps them honest, it makes them work.'

Ian and Beth are keen to see their staff get ahead financially. Both have seen too many shearers earn well all their lives but have nothing to show for it at the end. Ian comments, 'A lot of these Māori kids, they've had nothing in their lives, and they've become workers: they're making a lot of money and they're enjoying it. From our point of view we'd say, the sooner you get into your own house, get a little bit of debt for yourself, you'll win at the other end.'

What Ian and Beth have created is whānau. Beth comments:

'All the young ones that Ian has taught, they're like our extended whānau. Everyone seems to get on, though we have like twenty shearers and twenty woolhandlers; and they are all local and they've been with us for a while now. It's a good environment to be in. Sure, we have our ups and downs with our workers, but we enjoy it and we work together. And I reckon that's helped us be in it this long and make it successful for us. Cos it is a lot of work . . . They didn't want to be at school; a lot of them have been in trouble, and so they've come to us and Ian's said he'll give them a go.'

Ian adds, 'I just love it, and I love being able to give to these young boys.'

The shearing industry is a work environment where there are opportunities for those who want to grab them. Alistair and Kim Cuming are another couple who saw an opening and were willing to do the hard yards to success. Although he lacked an early background in shearing, Alistair took up an opportunity to shear while still young. And with his drive and a good head for business, he developed Rewa Shearing in the Rangitīkei from a small run into an operation with eight gangs running at peak times. Alistair and Kim talk through the complexities of contracting.

Alistair and Kim Cuming – Rewa Shearing

'I suppose we thought we were little hicks from Rewa – you know, we were just farmers' sons shearing sheep – and I guess we went to other jobs . . . we went back to farming, picking up hay, milking cows, working on a sheep farm, compared to these guys who shore sheep around New Zealand . . .'

The most that Alistair, a dairy farmer's son, had had to do with shearing was a couple of summers rousieing in a shed in Halcombe and watching his father shear the killers at home. 'I think the cutter and combs were pretty rusty and they only came once a year out of the tin where they had been put the year before.' In 1980–81 Barry Coleman, a mate, roped a seventeen-year-old Alistair into heading out with him to crutch some mobs for local farmers, so, with little shearing knowledge and a whole lot of enthusiasm, off he went. Barry was a hay contractor, and the shearing work came from his clients and friends. They shore in the little local two- and three-stand sheds with anything from a couple of hundred sheep through to about a thousand.

'We would go and crutch for a start, and you crutch sheep in the winter ready for lambing, or in the spring – early spring/late winter – and then we would go and shear them in summer after they'd lambed, and shear the lambs after that. So they were just acquaintances of ours, of mine or Barry's; just people in the district who would have seen two guys and thought, instead of getting shearers from Feilding we'll get these two.'

Alistair honed his shearing skills through Wool Board shearing schools, and when he needed more sheep to shear he travelled. He spent time with contractor Don Davies in North Canterbury, and did a couple of stints in Australia. He quickly built up his tallies.

'I think it was your first three hundred on lambs, which you would normally do first, and then your first three hundred on ewes. And then you would do your first four hundred on lambs and your first four hundred on ewes; that was in eight hours, or it was nine hours – I think I got up to 434 lambs in nine hours and that's as good as I got.

'I haven't shorn in a shearing shed for twenty years, but it's ingrained in your head because you'd done it so many times. And you're always looking for the perfect sheep and to do it the perfect way – you very rarely did it, but you wanted a quick last side because that's how you could catch up. You could be a blow in front of the guy shearing beside you or behind you. I don't know what it was like for other shearers, but to me it was a race. You didn't stop; you were sweating from the third sheep, and you just didn't stop till you'd finished the run – it was amazing . . . I didn't have a tally counter, but you just swung into the next sheep and you knew – that was the other amazing thing – you knew whether you were two in front or one behind or five in front because you worked everything out on the quarter: you'd got one sheep around them every quarter, the guy beside you; or you were one behind him every quarter, or one sheep behind him for the hour. You knew what you had to make up to get in front of him again.'

Alistair explains what happens when there's a problem sheep:

'Internally you are swearing at him. You might belt the sheep if you're an idiot, but you're trying to calm yourself down, calm the sheep down, get him shorn, get him out and then get the next one – like, get rid of that sheep cos he's cost you. So you might go into your pen – I called it, you'd farm your pen – and you'd always try and do that sticky sheep. You got to know what a sticky sheep was – you knew if he had yellow wool and all the rest were white, or his body condition . . . so that was the sheep you caught on the bell, so to speak: you had as long as you liked to shear him. A lot of that goes on: you tried to catch them on the hour because the shearer next to you might have stopped on the hour, so you would shear that through the hour while he changed his comb and cutter. You'd have a quick change – it was like a race car changing tyres, you would change your comb and cutter. He'd had a rest and you hadn't, but at least you'd caught up that minute and then you're away again. So, it was pressure every day . . . you just go for it . . . So you've shorn in your own little environment, and then you get asked to go and shear for the gang down the road and you've come up against their number one and away you go. It's all on – you can tell everyone's watching, everyone's seeing this little battle going on for the day – it's amazing.'

Alongside Alistair's tally-building was the development of Rewa Shearing, formed in 1984.

'The business started after a season of shearing for Jim Bull and realising that I had done all the organising with a bunch of friends and hadn't charged anything! So, basically I decided to start charging a contracting fee, and from there the word spread and the work evolved.'

In that same first year, a young woman, Kim Kellick, picked up some shed work at Rewa Shearing. From a farm herself, she was familiar with the inside of a woolshed but had never rousied before. She remembers the time as being 'fun; there was always a lot of humour and the shearers would have you on . . . oh, you know, put the handpiece on your bum and that sort of thing and make you jump. If you dropped your broom you had to shout a dozen beers that night; all those sorts of woolshed tricks that happen.' Somewhere among the woolshed fun Kim and Alistair struck up a friendship that led to marriage in 1989, and subsequently Kim became an integral part of Rewa Shearing.

Alistair had the drive to develop the business.

'I didn't buy my run off anyone. I built it up by getting offered sheds in the district, and away we went. I think I struck a time when older contractors were either retiring or getting to the end of their time in the industry, and farmers looked at me and saw a younger guy and thought, we'll give this guy a go. So, I didn't pick up twenty sheds off one contractor; I got, say, three off a lot of contractors and suddenly I had a run going by 1986 or '87.'

The nucleus of the gang was Alistair, Barry and Craig Northcote, a mate who was doing casual work in the district.

'Then we got my sister and her friend to rousie, so we suddenly had a gang, and then you were away. I don't think I approached farmers, but you certainly let them know you were there, and they would have known about you, and they were offering you their shearing and you were in business. It went from zero to a lot of sheds over three years pretty quickly.'

Jim Bull's shed had been gained through Jim's son, Adrian, who was shearing alongside Alistair at the time. It was a big stepping stone – there were 20,000 lambs to shear over January – but Alistair took up the challenge.

'You never said no to the shearing, you worried about the logistics later. You just planned it in your diaries and flew by the seat of your pants – cos at that time you're playing rugby, you're round the pub, with lots of young people around. You could always get people who had left school, or were on university holidays; lots of girlfriends and boyfriends as such would come and work for you in the holidays and the shearing was done in November, January and February, so you had plenty of staff and could grow it pretty quickly.'

Essentially what Alistair was building was a 'good run'.

'We had a long run, so you could come and start with me at the beginning of October with hoggets and get work right up until the fifth of January, right at the start, and that was a good run in those days. And I guess I had a reputation – you didn't get any time off: I'd work every weekend. Shearers liked to be known for getting a good run; having a good run with a contractor was the main thing.'

Changes in seasonal shearing practices had also acted in Alistair's favour as the business developed. Alistair explains the traditional shearing pattern:

'From the tenth of November to the fifth of January was the main shear. Second shear was from the first of March until the tenth of April, then you did crutching at, say, the end of July for four weeks until the last week in August. Then hoggets started in October, then ewes in November, then lambs in December. There was only five months between main shear and second shear, but those five months were the best growing months . . . they're growing wool flat out.'

The introduction of pre-lamb shearing in the North Island altered not only the shearing seasons but also the way in which Alistair ran the business.

'I don't know what year it all changed, but the first big change was that the ewes started getting shorn in the winter . . . it was a pre-lamb, so you shore them in June/July and then you shore them again in either December or January. So the seasons changed a bit and the crutching disappeared – you didn't have to crutch them because they'd been shorn. Then you were shearing ewes and lambs in January, which really put a lot of pressure on because the farmer didn't want to mismother the ewes and lambs by shearing them in November and December: he just left everything until January and then brought them in, to draft them, shear the lambs that didn't go to the works, then wean and then shear the ewes.

'Where I came from, all the shearers' quarters and sheds – a lot of them were fine for shearing in November/December when it was fine, but suddenly we were there shearing in June/July or July/August – they weren't set up for it. And that's when we went to not camping out as much; we started travelling to the sheds and went from a five o'clock start to a seven o'clock start.'

This was coupled with more of the shearers being disinclined to stay on camp-outs.

'When you're married, you'd rather be home; you'd rather sit in a car for an hour and a quarter, or a van, and be home with your family at night. And if it rained, as a contractor I didn't want them up there sitting in the shed for three days waiting for dry sheep, paying the cook, and eating food, so I'd pull them out. And then as well, the quarters got run down – the farmers weren't keeping the quarters up to scratch, so we weren't going to stay in a boar's nest . . . A lot of shearers used to shear for me in March/April and then go down south and do a pre-lamb. Well, then they didn't have to go down south – these married guys could stay in the district – and suddenly out of twenty-four shearers you had fourteen that could work all the year through with me . . . so you could basically offer people full-time employment.'

A real coup for Alistair, which further enhanced that 'good run', was when he was asked to take on the iconic Otairi Station in the Rangitīkei.

'I don't know whether it's a feather in your cap, but it's a big start to a young guy starting out to have a shed that's got ten stands. Everyone would soon know that you were shearing Otairi, and that filters down and helps you attract shearers again . . . They want to come to Otairi cos they get to shear for twenty-four days at the same shed. In those days they shore ewes and lambs on the first week in November, so the lambs were tiny and they could do plenty – shear good tallies – which attracted good shearers. It was a big thing.'

Alistair's first impressions of Otairi are imprinted in his memory:

'I remember driving down the first time and being blown away by the tarsealed driveway into a woolshed; I couldn't believe that. And the trees down the driveway, and everything painted new and tickety-boo. There were shepherds and horses and dogs everywhere, groundsmen, a big cookhouse that was busy with a husband-and-wife team doing that. There was an old Māori guy who drove the tractor with mattresses on the back and he'd go to the shed – when you shifted from the front shed to the back shed. It was a busy place . . . I think we were doing 18,000 ewes, so 18,000 lambs – they were probably 100 per cent lambing in those days, 105 per cent.'

Other big stations came on board after Alistair picked up Otairi. Te Kumu, Hurleys' Papanui, the Batley sheds up at Taihape with 'seven and six thousand ewes – they were big sheds'. Ferndale came on board later in 2008 or 2010. There was some diplomacy involved; Alistair recalls how 'some of those big stations didn't want you to have their shearing cos they thought they'd be second fiddle to Otairi – so you had to let them know you could handle it – you had to be big enough'.

Logistically, of course, the growth meant more juggling of staff.

ABOVE A typical scene at shearing time, a line-up of vehicles outside the woolshed, Otairi Station.

LEFT A young woolhandler clearing the fleece away from the shearer. Keeping the board clear for the shearer means the woolhandler has to be deft with her paddle. A shearer may not be quite so forgiving if a woolhandler consistently gets in the way or bumps them.

'I'd get my diary: I'd make sure I had twenty-four shearers and twenty-four rousies, and then I had lots of other friends that had shearing gangs, so you'd ring them up and maybe get a shearer if you needed one for your other sheds, so you could make sure you had enough at Otairi. You always looked after those guys, but then you also had other sheds that were just as important to do, so you were immersed – you made sure you had plenty of people.'

An aspect of the growth was dealing with camp-out sheds. Initially the crew took their own vehicles out; and then, as Kim remembers, there were the logistics of sorting the cook and the food:

'We had Cortina station wagons and I'd go into Write Price and I'd get three trolley-loads of food, set the cook up. We had all the pots and pans and everything and we'd fill up the Cortina to the absolute max and send her off on a Sunday night. And then probably about five or six days later you might get a phone call or a message that she was out of flour, butter, whatever, so there would be a run – the rural delivery run would drop off whatever, if it was out at Otairi or Alistair would be going up, and you'd just restock her the whole time cos she could be away for up to six weeks.'

Summer was the busy time for stay-outs. Logistically it was easier; there wasn't the intense daily planning required.

'Cos once you got them to Otairi there was a whole eight shearers, eight rousies and two pressers all on their own with the ganger looking after them, if all went well. Unless the manager would ring up and say, you're to take such-and-such out of the shed, or the rousies or the ganger would ring up and say could you take such-and-such home, he's sick or crook . . . but once they were away they all looked after themselves, which made it a lot easier.'

Due to the influx of 'foreigners' that Alistair needed to bring in to cope with the growth, further infrastructure was needed. The 'Pig Pen' and the 'Rabbit Hutch' were little shearers' quarters that the business had in Rewa; imports could flat there.

'You were just zoned in. You'd tell your shearers when they'd go down south, "See if you can bring a couple of shearers back with you." And they were good, they'd do that . . . they wouldn't bring a ratbag. They knew what we wanted, they'd bring in good blokes. And there weren't many ratbags in the industry, really. As one of my shearers said, there's only three weeks' difference between a good haircut and a bad one, so it didn't matter much.'

With today's cellphones, it's easy to forget that contractors had to factor in less-than-convenient communication methods in 'the old days'. Landlines were the only option and contractors would use farmers' phones – or, recalls Alistair, 'Sometimes I'd have to get in the car and go to the pub, go to the Ōhingaiti or wherever to find someone . . .' Kim, at home with the children, remembers the phone always being in use:

'Also in those days we had no call waiting or anything like that to begin with. So your phone would be engaged for hours and hours and hours, and for someone trying to ring – like, for a farmer to ring in to us – it was a nightmare for them because they could never get in. Our phone would start at five in the morning and finish at half past ten at night.'

When call waiting was introduced, Kim remembers it being a 'gift'. Alistair remembers it speeding things up: 'When you flicked over you could say, "You got your sheep in, Bill?" "Yip" – "Good as gold" – *bang*. And that's done. That's all you needed to know.'

The next technological advancement in communication was the cellphone. Alistair laughs with the memory of his first one. He took up this new technology very early, remembering it being the equivalent size of his A4 diary:

'I can remember walking around the saleyards thinking I was God's gift to communication because I suddenly had this cellphone . . . no one else had them, the shearers didn't have them . . . but at least you could cart your phone around with you.'

While cellphones have undoubtedly made the contractor's job much easier, there's still a logistical maze to negotiate to make sure everyone is where they should be. Kim remembers it being like a 'jigsaw puzzle'. Alistair explains:

'You wanted them back to the same shed, but I didn't want one gang finishing at three o'clock and another one not finishing, so you might have to take one shearer out of that shed that was fast to go to the shed that wasn't going to finish, and put the slow shearer into those guys so that they would then finish at four o'clock and the other gang would cut their shed out. So, there was all that – and probably some contractors may not have been bothered with that, but I definitely was. You wanted to finish that shed and that gave everyone a longer day. You expected your shearers to go past five o'clock to finish a shed, and you got into a good culture in your gang – they would do that; and you would shout them a beer, and that's just how it went.'

Alistair remembers a multitude of variables all influencing the decision-making process:

'The weather; then farmers' tallies (they don't have their correct tally); the contractors don't get the right shearers; and then the sheep – when you do start shearing and you think that gang can do a thousand a day and suddenly they're sticky and they do eight hundred, or at three o'clock the farmer's about to get the last two hundred in and it rains, he gets caught in a shower shifting them from one shed to another. There's all these things . . . and suddenly at three o'clock my ganger would ring up and tell the farmer, "You've got to ring Alistair cos we're not going to cut the shed out." And then it would be mayhem because the next farmer had his sheep in and you had to ring him up and he'd say, "What the hell happened to Bob?" "Bob got caught." "Well, fuck Bob, I've got a thousand in for tomorrow." And I'd say, "You'll have to let two hundred go cos we're not going to start till smoko." So there were all those different things going on.'

Added to that are all the interrelationships that a contractor must grapple with. Alistair looks back reflectively over his years:

'It's probably the same in any business when I look at it now. You've got your big clients and you've got your grumpy clients, and you've got the ones you can tell them to let the sheep out and they don't mind and there's the ones you know do mind; and you play on all those different things. There's farmers you can tell we're a week late, and there are farmers who would have you on about it: "You can't be a week late, it's been fine."

'And you're dealing with people. I used to say, they're human, they're people, they sleep in, they have a bad night, they have an argument with their wife, or the girlfriend storms off and she's not around to go to work in the morning. All I'd say to my shearers is, whatever happens, just let me know. I don't care if you're not going to turn up for work, but let me know at four o'clock in the morning not at six, cos at four o'clock I can think about it and manoeuvre it all around – take a rousie out and put someone in there, or ring somebody and say, "We're stuck, can you come and work?" But when they ring you at six or seven o'clock or they don't ring you at all and the ganger rings up and says, "I'm waiting outside Michelle's door and there's no one – they're not here," then you've got to reconfigure again during the day.

'I've had shearers cut tits off hoggets – so that's your young replacement sheep – and not raddle them, so we've lost sheds over that. I used to say to farmers, "He's had a bad day – I know he's going through trouble with his wife or his family at home," or "He's got temperament issues and he won't ever come back to your shed." Sometimes that worked, and the farmer would forgive you. Or other times, I've bought those sheep, paid for them, and you still kept the shed. Other times the farmer's said, "That's enough" and you're gone. It's not your fault, it's just one guy in there . . .'

Kim adds, 'So it's Rewa Shearing that cost us the shed – it was never Joe Bloggs – you just bore that.'

While no contractor likes to lose a shed, there are times where it does happen.

'You lost sheds when farms changed hands, you lost sheds when a different manager came to run a farm, so he had his pet shearers or contractor; you lost sheds if you did a bad job. We lost sheds because the farmer said we did a bad job with wool. There's nothing you can do; you just say you're sorry, you did your best, and you move on. You try not to end on a bad note with him, because I've lost sheds and got them back three years later, five years later – so you grin and bear it.'

It's got to be said, though, that there were swings and roundabouts.

'I remember being at a station and I was outside counting out with the farm manager, and he was going crook at this shearer's sheep; and the shearer pushed open the window looking out over the pens and said, "Alistair, you can't make a silk purse out of a sow's ear," then quickly closed the window and left me to talk to the manager. I think he might have just about smiled. If the sheep were well presented, the farmer had a watertight case for them to be shorn well; but if they were skinny and not in good order then it was on our side – you could say, "Look, these are hard to shear. They're doing the best they can, but they can't do a good job out of them."'

Alistair comments on the need for humour in the shed: 'You had to have a joker in the gang . . . because you had to have fun. Because it's a long day without laughing, and so you wanted good fun people in the gang.' He remembers his mate Barry Coleman as being a real joker:

'Barry would be waiting in the pen when you'd finished a sheep and you'd walk in to grab a sheep, and he'd jump on you and there would be an almighty fight in the catching pen. Or he'd be play-fighting with the shepherds – that was when big-time wrestling was a big thing – and we'd all be jumping up and down on the shepherd. And then they would go and get three of their shepherds – and that was not a fight, it was just fun.'

Rousies were fair game, too.

'You could be having the rousies on to go and ask the farmer for white raddle. They'd come back in very red-faced after they'd walked all the way through the yards, and the farmer would tell them – he'd play along with it. You'd get the rousies to raddle a sheep's pizzle because the farmer wanted to count the pizzles – you could have a lot of fun.'

As the business grew, Alistair stopped shearing after discovering it was too difficult to fill in. 'The day you went and filled in was when everything fell apart. If you had to go and help do half a day somewhere, you would have been better home on the phone organising.' There was some regret with not having time for shearing, but Alistair was pragmatic. 'I'd liked my shearing, but I also knew that I was making money. By then I'd realised I could make money out of being a contractor, so I was more interested in making sure that I had sheep in, and shearers and rousies and the cook and everything was organised.'

Sitting in their home in Havelock North reminiscing, it seems a very long time since Alistair was starting out, rushing from milking to shearing. He remembers the long days:

'I would get up in the morning and help Dad milk the cows at probably five o'clock, half past – we milked about two hundred cows. Dad would let me out early, so I'd leave the cowshed at seven for a quick breakfast, and we used to start shearing at eight. Shear until five o'clock – or if they were little sheds you might be finished early – and then shoot home and help Dad finish off milking the cows, because he wouldn't be finished until probably quarter to six. Hose down the cowshed and feed the calves or whatever else had to be done, and then dinner with Mum and then to bed.'

Alistair and Kim ran Rewa Shearing from 1984 until 2004, when they brought managers in. In 2016 they sold it to Tracey and Glen Thomas, who have kept on the Cumings' manager, David Gordon. Alistair had grown the run to incorporate sheds in a 100-kilometre radius from Hunterville, Bulls, Feilding, Colyton, Taihape and out to Papanui Station, running eight gangs at peak times with seventy to 100 staff. Not bad for a 'hick' from Rewa.

In sheds around the country, at any given time, there are shearing gangs working away to ensure the nation's flock is decloaked. Within the shed each person has a very specific role; completing the job successfully requires teamwork, focus and a lot of good humour. It is a place of opportunity for those who are motivated to excel. The role of the shearer is obviously central. Theirs is a repetitive task that requires strength, skill, mental stamina and a huge amount of guts. As Parky says, 'Oh, it's not an easy job – be one of the hardest jobs in the world, I would think, to be a professional shearer.'

Every day, shearers travel out to sheds around the country. Not all travel, however, is created equal. The road in to Muzzle Station, between the Seaward and Inland Kaikōura ranges, is at the extreme end, taking two hours to travel 30 kilometres. Sometimes shearers may even be flown in to this station if the track is not negotiable.

3

Their World

resser Kevin Peek with a
mountain of merino wool to
ress. Quail Flat, Muzzle Station.

Every job has its own associated culture, and the shearing industry is no exception. How can it not, when teams of colourful characters load themselves into vehicles each day to travel some of the toughest roads in the country, and then take on the chin whatever conditions are served up to them at the woolshed (and accommodation when on stay-out), while doing one of the toughest jobs around?

Humour is essential in helping to break the intensity of the atmosphere. Everyone's idea of humour can, of course, vary, and just how far the envelope can be pushed depends on the culture of the particular gang. While some antics are best left unwritten, those that are common to many gangs are nailing moccasins to the board, locking the pen door to delay the shearer's exit or wrapping the start cord up and around the machine. They're all light-hearted tactics to cheat and get ahead of your mate on the next-door stand. A less common trick might be to tie a live possum to the belly of a sheep to surprise some poor unsuspecting shearer.

Music, too, has long been a backdrop to the work in the shed, and when chatting with a shearer the era they worked in can be gauged by the music they talk of, from Elvis or ELO to a bit of Bob Marley or heavy metal. The older workers always complain about the young ones' taste, although the most unusual example must be the old guy that Reg Benjamin remembers bringing his record player – its needle skipping to the shed vibrations – so he could play Mario Lanza LPs as the shearers pumped out the sheep.

Back in the day, a large part of the shearing culture evolved through the stay-out environment. With teams working and living together, often for long stretches as they travelled from shed to shed, not going home unless it rained, strong bonds developed. The older shearers all talked of the camaraderie on stay-outs, recalling those days with nostalgia. The gangs became like family, often with a cook – whose output could make or break a gang – acting like a camp mother to the younger members. These days it's far more common for shearing crews to find themselves in the back of a van travelling to a shed daily than it is to be staying out on location. Of course, some sheds will always remain as stay-outs, their

ABOVE Loud music is sure to be heard in woolsheds around the country.

LEFT Suzanne Webby cooking for the Pullin team at Quail Flat. A shearing cook learns to take any kitchen conditions in their stride as they churn out vast quantities of food to fuel all those working in the shed.

remoteness making the logistics of a daily commute impractical. For some shearers today they are an enjoyable novelty.

With good reason, the maxim 'work hard, play hard' is commonly heard about those in the shearing industry. For some it might be fair to say that the maxim was 'work hard, play harder'. Shearing produces a thirst, and slaking it with a cold beer at the end of the day has long been etched into the shearers' psyche. And while the 'play hard' culture has toned down over the years, it's fair to say the breweries have done well out of the shearing community, and there would be a few hotels around the country that owed their survival to it. Malcolm Sadler remembers the eight-stand gang at Waipuna drinking the Ikamatua pub dry one year. Prior to the change in attitude and regulation around drink-driving, it was nothing for a crew to turn up to the pub on their way home for a couple of jugs and for that to turn into a five-hour session. Kelly Hokianga thought there would often be more sheep shorn in the pub than in the shed. And yet there has always been an unwritten code among shearers that if you played hard at night, you still turned up the next day, and you worked as hard as – or even harder than – the day before. There was no tolerance of shirking if you wanted to keep your job. Some of the older shearers might just say that it's one code broken more frequently by the younger ones today.

The travel, the work, the stay-outs, the food, the drink, the music, the sense of family: all have been part of the blend of the shearing culture for decades. Each generation of shearers may have slightly different experiences of them, but together they speak to the colour and vibrancy of a large and significant community within the broader pastoral industry.

Belonging to that vibrant and colourful community for over 100 years is the Hape family, contracting in the Wairarapa. The Māori whānau shearing gangs are the likely early model for the modern contracting businesses, and the Hape family share their memories of what it is to be part of a whānau gang.

TOP One of the joys of sleepouts are opportunities for hunting in remote locations. Here some of the Pullin team at Quail Flat search for chamois on the hills.

BOTTOM Even at remote locations like Mt Nicholas on the shores of Lake Wakatipu the impact of technology can still be noticed.

Hape Shearing –
Whānau shearing

It's difficult to fully capture in words the pride, joy and enthusiasm expressed as three families of Hapes recounted their shearing history. There were three separate interviews. There was Molly Hape, and her children, Tia, Melodie, Pat and Robbie, sitting around the table at Robbie's home in Dannevirke, where there was a lot of laughter and teasing as they talked. There was eighty-nine-year-old Ivan Hape, grand-uncle of Molly's four, interviewed in the living room of his home out on his farm on the outskirts of Dannevirke, with his wife Miriama pottering around the kitchen in the background. Then there was Hayden Hape and his wife Stacy, who took time out on their way home to Dannevirke to meet in a quiet corner in a Pahīatua pub. They are all proud of being part of a long tradition of Māori contract shearing, and providing for many families in Dannevirke – as well as having a whakapapa that takes them back to legends Johnny Hape, Woppi Hape and Ihakara Te Tuku Rapana (aka Ike Robin), a highly respected businessman, orator, philanthropist, wrestler and gun shearer, with a top tally of 358. These three men excelled in shearing and left their mark – not only with their tallies but also, more importantly, with the skills they passed on to their tamariki and mokopuna.

Listening to the three Hape generations, what becomes apparent is how shearing gets under the skin, or perhaps flows through one's blood. Ivan, born in 1928, has the longest memory, which begins with him eager to join his father, Johnny Hape, in the shed:

'But I can remember when my dad at the evening meal would say to Mother that he's going out to one of the sheds to see how they're going. And I would be out, first thing in the morning, sitting in a car waiting for him to take off. I spent quite a bit of my youth before I went to school out in the woolsheds.'

He recalls that safety in the shed was an issue, so he 'was only allowed to move around in the press room. Children were never allowed up on the shearing board because of the sheep and the shearing machines.' There was much for a young lad to watch.

'Everybody was moving around, the rouseabouts were carrying fleeces and there was a general hive of activity . . . with those old types of machinery that was in the sheds in those early days there was that much noise if you wanted to be heard you'd have to yell and scream.'

According to Ivan it was Johnny's father, Bob Hape, who started the run in the 1890s, but he knows little of Bob's time other than the stories his father told him. Johnny had to start off as a rouseabout for his father before he took up the handpiece. As Johnny worked, Ivan relates how 'he said he would walk along the board or he would watch each shearer and if they were quicker than the others on a certain part of the sheep he would concentrate and learn all the blows on that particular part of the sheep and then he would move on . . . that's how he told us how he learned to shear. He was self-taught and he became quite a good shearer.'

'Quite a good shearer' is, of course, an understatement. AR Mills, author of *Sheep-O! The story of the world's fastest shearers*, travelled much of the country and saw many of the best shearers of his day at work. Mills decreed:

> *Johnny Hape was an outstanding shearer, and he was the yardstick by which I measured all others. If ever a shearer got his sheep easily it was Johnny – always the same number of steps in and out of the catching pen. He shore with methodical precision – the same number of blows to each sheep – and his work was invariably clean, an object lesson to all in the shed with him.*
>
> *He had been schooled by his father, who held him down to a daily tally of 150 for several seasons, until one morning he said, 'Right Johnny, you can open out now.'*
>
> *Johnny did just that. He immediately 'rang the shed' and was never beaten from that morning until the day he knocked off shearing. He never became a world record holder, but then he never made an attempt.*

Mills went so far as to state that 'the late Johnny Hape, amongst a galaxy of talent, stood out as a shining star in the shearing firmament'.

Among Ivan's shearing memorabilia is a small newspaper cutting from the *Dannevirke Evening Post* that also speaks to Johnny's shearing prowess. Having already put up a record-breaking tally of 328 Lincoln ewes, when that was beaten he set to and re-established his dominance by shearing 361 lambs in nine hours at the Waitahora shed in January 1919. The handpiece, a Wolseley No. 6 presented to Johnny at the time of the record, is still in Ivan's safe keeping.

Ivan believes Johnny took over Bob's contracting business around 1920, and from there he worked on building up a larger, very successful run. During World War II the size of the business reduced due to petrol rationing and the subsequent restrictions on travel. Johnny also lost some of his gang when they enlisted. Ivan states that his father argued against their enlistment as shearing was an essential service but that, despite dispensation, some of the men went anyway.

Ivan did not take on shearing until 1948, when he was nineteen, but his training had started with fly crutching out at Akitio. He laughs as he remembers realising at the end of the day's work that the rousies had earned more than him:

'So I said to my dad, "This is not working out – can I go back to school?" He said, "No, no, you wanted to go to work." He said, "You'll be working for the rest of your life." And I was. When I first started to shear, because of the closeness of all the sheds to where my dad's farm was, I used to shear all day, nine hours – shear in a woolshed for nine hours and come home and do agricultural work till eleven, then I'd go back to bed and be up at four o'clock in the morning. I used to love it. I loved shearing.'

Johnny's teaching style was similar to his father's. He limited the young learners, of which there were always one or two in each shed, to a certain number of sheep per day.

'When you learned to shear for Johnny Hape you weren't allowed to do any more than a hundred. Once you did a hundred, that was your first twelve months, and the shearing season consists of three months, and you were never allowed during those three months to shear more than a hundred. Your second shear out – and I'm only talking shearing your second year out – you were allowed to do 160 after that. If you shore above your tally – if you shore twenty or thirty more than you were allowed to – you were kicked off the shearing board and put back to rouseabout, and you were there until you learned that you couldn't do that.'

Johnny's approach instilled in Ivan best practice, and those lessons stick with Ivan today.

'If I can say this about shearing, it is the hardest, hardest work that I've ever done. If you do not learn how to shear properly, it's real hard work; you're taught to sit the sheep in a proper place and the sheep won't move, wriggle, it will just sit there. Unless you're taught where to position the sheep you are battling the sheep the whole time you are shearing – it's moving around, wriggling about, and this is all part of learning how to shear, and I was an apprentice for six years.

And after six years I shore my first three hundred, and my dad says to me, "Right, you're away."'

Ivan's best tallies were around the 320 mark, and he emphasises they were clean:

'My sheep, you couldn't see where it was started or finished; they were clean, and they were like that every day. If you're taught to shear properly right from the word go, it's just a natural instinct when you pick up a sheep and you shear it; your handpiece goes into the groove.'

The gangs were divided into two runs. Ivan's cousin Woppi had the Burnview run out at Herbertville, taking in Delby Holden's station at Raumati, Burnview Station and into Waewaepa. Ivan shore in his father's shearing run, which started at Waipawa, through Delby Holden's station at Tikokino, then on to Glenora, a Frank Armstrong property, then on to Kaitoke Station just out of Dannevirke. These gangs consisted of eight shearers.

'For every shearer there was supposed to be a rouseabout – so there were eight rouseabouts, two pressers, an expert . . . the expert ground your shearing gear and if there was a breakdown in the gear he was an expert – he did all the repairing. And then you had a sheepo. I suppose we would be close to twenty-five including the cook, and sometimes there was a cook's assistant.'

They shore at each station twice in the season: first they shore the dry stock – the hoggets, rams and dry ewes – then they would go back to do the wet run, the ewes with their lambs. In between, the two gangs were divided up into smaller gangs to shear around the smaller sheds in the district. At the time their runs only shore full-fleece wool and the season was still short, September to January. Ivan remembers with a smile how there was a strong competitive streak among them all: 'Each gang would try and beat the other, so they would get one of the other gang's sheds. Which never eventuated.'

Eventually Ivan was given the role of pannikin boss (ganger or team leader in today's parlance), and his run was within a ten-mile radius of Dannevirke.

'As the pannikin boss you had to get the best you could for the shearing gang; you also had to make sure that the gang were doing a good job for the owner, and that entailed making sure the rouseabouts did the work around the table properly and the shearers were shearing the sheep to the best of their ability. So really the pannikin boss was the go-between the shearers and the owner. He was the kicking box for the gang and the kicking box for the owner.'

Ivan laughs with the memory of those times and comments how he enjoyed having the responsibility. And he can still rattle off the menu from that era – which is not surprising, given the simplicity of the diet:

'The main item for breakfast was potatoes and chops. You always got a mutton a day if you were a gang of eight, eight shearers. There was always a mutton supplied by the owner. And if you were a gang of five – if there were only five shearers – you would get a mutton every second day. But for a full board there was a mutton supplied by the owner every day . . . Then the other food was supplied by the contractor, he'd get your greens. If you had porridge, you'd have porridge; but the staple diet was meat and potatoes for breakfast, meat and potatoes for lunch, and meat and potatoes for tea. There were no side things, such as puddings or Creamoata for breakfast – no Skippy cornflakes. Smokos were sandwiches and biscuits. It was just mostly meat sandwiches . . . more mutton. And sometimes the contractor would throw in water biscuits.'

Ivan laughs as the memories swirl back into focus:

'We got plenty of those. A dry cracker – they called that the Hape special. Oh, that brings back memories. There were no cakes, no sweets. Mind you, when you're a shearer you don't look forward to anything like that – potato and meat were the staple diet and that's all they looked for . . . sometimes they had stew for tea . . . yeah, they had boil-ups. Mostly boil-ups for their evening meal.'

Ivan remembers a family-oriented business. Families worked closely together while staying out on the stations where they shore:

'Most of the Māori families around Dannevirke were all related, and whenever they went out shearing it was just one family that went out. You would have a few outsiders that were employed by my father to either shear or rouseabout. When I was involved in shearing there were hardly any families with their children. If there were children in the family the mother would stop home with the kids – it was just the man who would go out shearing.'

Ivan recalls the rousies were women, either single or with children who were old enough to look after themselves:

'Yeah, shearing was a good life. Shearing was a life where I got to meet my family . . . I actually ran a gang for my dad and it consisted of five shearers, five rouseabouts and a couple of pressers, and that was a dozen people. And for three months we were one whole family. If you were sick, everybody got sick with you.

'I thoroughly enjoyed my life as a shearer, because you'd never ever strike a job like that where it was family-oriented – you were one great big family. You didn't have to worry about your food: it was always there. You went to work at five o'clock, you finished at five, you had a couple of beers and talked with your mates.'

Beer, Ivan recalls, was drunk in moderation, and if there was a party it was Saturday night, as they had Sundays off unless it had been a bad season; then they would work every day of the week. Music was, of course, a feature, and Ivan still remembers Dereck Kara and Bill Hetariki with their guitars. Sunday generally was a day of rest, other than perhaps doing their washing, and attending church if they wanted to.

Although he enjoyed shearing, Ivan did not stay in the industry.

'My dad, when I first started, when he first taught me how to shear, said to me I'd be only good for twelve, thirteen, fourteen years. I asked him why, and he said to me that I would be taking the place of another young fella coming up and that I should stand aside. So I shore for fifteen years and I said to my dad, "This is my last year." And he said to me, "Oh, you've still got a few years." And I said, "No, when you first taught me to shear you said I was only going to be good enough for X amount of time and I've finished."'

Ivan laughs at the memory of that conversation, but also ruefully remembers that heading into sheep farming was no picnic. With 1200 ewes and initially no fencing, for the first six months he chased sheep off the roads until he could afford fencing. And the request for a tractor was met with a simple reply from his father:

'I said to my dad, "Can I have a tractor?" And he said, "When you've got the money to buy yourself a tractor you can." So I used to wheelbarrow my hay bales out to the cattle, and I did this in the dark so the neighbours wouldn't see me.'

Ivan is uncertain when his father stopped running the contracting business, but he knows that when Johnny was satisfied that Woppi knew how to look after the business the mantle was passed to him. While Ivan went farming, Woppi ran the shearing contracting business, but the profits they made were put into land. Johnny had bought quite a few farms and dispersed them among his family, and Woppi continued to do the same.

As Ivan reminisced, there was a real sense of pride in his father's achievements and his high standards. A humble, self-taught businessman, Johnny placed emphasis on good clean shearing, as Ivan recalls: 'We had the honour of being the cleanest shearing gang in the district. We didn't know how to shear a sheep rough. We were taught to shear the sheep clean and every owner got a good cut.' Ivan slips in that Bill Meech and Allan Peeti were shearers who worked in Hape sheds before they took on contracting. 'Bill Meech turned into a real big contractor. And he started out with Woppi out at Burnview . . . little wee man – he was smaller than I was.' All those working in Johnny's gangs, Ivan recalls, were taught to shear properly and so as a result enjoyed their shearing. 'I've actually seen shearers who were never taught to shear properly, and it was hard work just watching.'

Woppi could not help but be a gun shearer, with blood ties to both Ike Robin and Johnny Hape. Having foundational training from Johnny, Woppi excelled in the shed, and he is perhaps best known for a record lamb shear. Again, AR Mills opines, 'The first really outstanding lamb tally of which I am aware, and one which should be given "official" status as a world record, was W. Hape's 422. These lambs came into the shed – to use a term which means without previous preparation – straight from the paddock. He caught all his own lambs from a double catching pen which measured 9 feet by 9 feet, and it was not kept full, either.' Such prowess built respect; according to grandson Hayden Hape:

'One story about our grandfather, Woppi, he always carried a letter with him challenging Godfrey Bowen, and Godfrey Bowen would never accept the challenge because the Queen was coming to New Zealand and if he had lost to our grandfather he wouldn't get the opportunity to meet her. So that's a story that was always told amongst the elderly ones that worked for our grandfather.'

Woppi passed away in 1969 and Johnny died shortly after. As Woppi's boys, Robert, Patrick and David, were too young to run the business, it went to Woppi's daughter. Eventually the boys took the reins, but by then the run was smaller as parts of the business had been sold off. One who remembers Woppi's era is Molly Hotereni, or Molly Hape after she married Woppi's son Robert.

Molly has many warm memories of those early days working for her soon-to-be father-in-law as a woolhandler. She became head rousie and taught new workers the ropes as they came into the shed.

'When I first got here, and we went into the shearing gang, 1966 was my first year here; I was still at high school. I wanted to come over here and make money to get myself my first scooter, to be the first Māori in Foxton High School to get a motorbike, and I did. It was my father-in-law that had the run then, but I didn't know he was going to be my father-in-law then. I was here with my sister and my brother-in-law and we were sitting in the house that my father-in-law lived in and he had these great big green trucks, he had three of them. The first truck come in – this was just before main shear, getting ready for main shear – and it had people from Raglan, Hamilton, and around that area, on the back of the truck and they just jumped off the back of the truck with their bags, heaps of them; and then all of a sudden that truck goes and another great big green truck would come in and they were from Foxton. My sister used to send a truck back there and pick up the Foxton crew and bring them over . . . and then another truck would come in and they would be from Rātana, the Rātana whānau . . . and that was the gang. They'd all come over and sit outside the big whare, the homestead. My father-in-law would go out there and put them into their gangs. They had about eight different gangs then, and all the – they called them pannikin bosses, they were the head of the gangs – they would get them and put them in different parts of the whare, and then he'd put all different people in those, in their gangs; and that's how they got their gangs going.'

Molly tries to recall how many people came in:

'Oh jingos, there was heaps – three truckloads of people. They had no canvas on the back to close them, and when you were riding along you could see everything

A Hape gang at Mangaorapa Station woolshed, Southern Hawkes Bay.

you were passing. They were the big Bedford trucks, green; they had no name on it, but everybody knew that they were the Hape trucks. They'd even send a truck up to Auckland and bring back all these people and they'd just fill all the quarters up. The whare was out at Tipapakuku, just out of town. He had about four rooms with double bunks and there was about eight double bunks in all the rooms, and then there was a couple for married people and another couple for the girls and boys and whoever cooks. We had our one main cook there and that was Aunty Molly [Woppi's half-sister].'

The whare, known as the 'big house', acted as a central base for operations. Molly remembers a big cookhouse where the cook fed everybody, as well as a big truck shed with a service pit where the trucks were maintained.

Most of the roads they travelled on were still metal, and Molly talks of her husband and his brothers, Patrick and Dave, learning to drive on those roads. She recalls the truck trips fondly, despite the road conditions being less than perfect:

'They were the bomb on those roads cos you slide like this, and they knew how to slide . . . Oh it was awesome, cos you got your beers on there and you got all your bags, and everything's all stacked up. You had a truck with all your gear on it . . . and another truck for the people, and we all used to throw mattresses in and sit in the back . . . we'd take about three and that'd fill the whole truck, and everybody would sit on those. We'd have a guitar . . . and people singing, and yeah, it was really neat.'

The musical backdrop Molly remembers included 'Ten Guitars', and Chubby Checker and Elvis songs.

While reiterating how much fun it was, Molly quickly launches into a story of mischief:

'I'll tell you a story about my uncle; he's passed now, he's a Nicholson. Woppi, my father-in-law, he never used to like them, cos there's a lot of pubs on the way out to the sheds, and he never liked the trucks to stop at any of those pubs . . . he also had buses to bus people . . . but my Uncle Tiwai, one day just before they were going past a pub he opened the window and threw his bag out the window, and everybody saw it. "Oh, bus driver! I've dropped my bag and I need it cos it's got my handpiece and everything in it, my shearing gear . . ." And he said, "Oh you effing —!" Because it had all his shearing gear in it they stopped, and the bus driver said,

"Okay, I'll go and get your bag." He opened the door, everybody was out, and they had to ring up my father-in-law to go and get them out of the pub. Oh, he wasn't a happy chappy. So, the shed that they went to, they were not allowed to take beer there. My father-in-law said, "No beer there – that's for doing that to me." They said, "Oh okay." So when he went, they used to go to the cocky's house – the farmer's house – and ring the pub to order some beer. Then they used to ring the taxi in Masterton, and the taxi driver would bring the crates out to the shed. Not right to the shed: there were creeks coming from the hills, he used to put them in there – about ten crates of beer . . . and that's how they used to get their beer.'

Molly herself may have made mischief, too. She laughs as she relates how they would harass the Ben Nevis big Drysdale rams, which required two to shear them:

'Us stupid things, at night we'd go over there when the rams were in there and we'd make them all wild and make them chase us until we got a growling. And they said, "Don't do that cos tomorrow they'll be still mad!" They'd say, "We'll get you guys to hold them." We thought it was fun, but it wasn't. We learned then never to do that.'

Undoubtedly guidance was needed for the younger members of the gang. 'Aunty Molly was the one: Molly Tawhai, Woppi's half-sister, she was the bomb cook. If you were new at the shed and you were single she'd keep you under her wing, but if you'd been there and done that before she'd make sure you were doing alright and nobody's hassling you, specially the boys.' While Aunty Molly may have acted as camp mother, Molly recalls the shearers being good with the pressers and the sheepos 'cos they were like teenagers – they were always young and smart, and know-it-alls, and they kept them in line; but there were a few rousies, too, that needed to be kept in line.' She recalls one head rousie, Dereck Kara, looking out for the younger ones, too. Despite the care of the older ones they could be tough as well. 'Some of the older ones were so hard; but I'm glad they were because anything they taught us it's just gone through the whānau and the workers.'

Robert, Patrick and Dave were the fourth generation of Hapes to run the contracting business, and a fifth generation came up behind them as their children were introduced at an early age to the shearing life. Robert's boys, Robbie and Pat, and Patrick's son Hayden share warm memories of childhoods spent in the sheds. Pat kicks off:

'I'd get up early when I was four or five – cold mornings – and jump into the Bedford vans going to work. There's Uncle Pat and Uncle Dave and Dad, the three gangs are going. We always used to run to the fastest uncle that could shear, so I'd jump in with Uncle Pat . . . (I'm named after him) in the flashest van – they had three Bedford vans . . . and we always used to sit up on the bonnet, it was nice and warm. It had a big box on the front by the steering wheel and we just used to sit up there . . . and go to sleep until we'd get to the shed and wake up and straight to the farm. "Oh, could we ride your bike today?" "What are we doing today, boss?" "Oh, we've gotta go out the back and get these sheep in." So, I only come back for lunch and smoko with the gang and watch all the boys shear; that was quite cool. And asking for turns, like doing the last side . . . just learning.'

There is laughter among the family when Pat mentions the inevitable tears if they missed their ride. Robbie reiterates, 'I've had the same sort of upbringing as Pat: we both raced for that front seat of the van – nice and warm – first in, first served . . .'

Pat goes on to tell of other childhood fun:

'We used to go eeling through the day, go crawlie hunting, bring the boys back a feed of little crawlies and eels. That's the fun we had when I was growing up round the shed. It was awesome – chasing turkeys, trying to kill a turkey for my uncles . . . things like that. When we used to go on stay-out, we used to play war in the shed. Grab a stick and a sock and tie it up, and smash all the raddle up in it, and every time we touched them we'd whack them and leave a big chalk mark on the back of their legs or on their ears sometimes. That was a lot of fun with the cousins and the young pressers; it was good fun.'

Cousin Hayden recalls memories from stay-outs as well as the inevitable pub visits:

'I suppose my memories of the woolsheds are old cold buildings and watching my mum in the kitchen. Walking out of a room all by myself because Dad and that have already gone to work. They start at four in the morning, four-thirty in the morning, start shearing at five. In some cases, if it was a twenty-five-minute walk to the woolshed from the quarters – say Glenora Station, it was a bit of a walk – they would leave early . . . I would wake up just before morning breakfast so I would see them all coming back. But sometimes I'd get up maybe six o'clock, and I'd get organised and

LEFT Sid Nicholson followed in his father's footsteps and joined the Hape shearing gang as a boy. Back then they shore full wool sheep between September until the end of January. On the off season he went scrub cutting. Having shorn for 52 years Sid has shorn around much of the country and defleeced 1.5 million sheep.

BOTTOM Taking a well-earned break for lunch, members of the Hape gang relax and enjoy a laugh together.

realise Pat and Rob had gone, and I'd go over to the woolshed and they'd be sitting in the pens watching. Their dad wasn't allowed to move without them.

'Every day was different, every season was different; you laugh when you see the kids sitting outside the pub and stuff on these movies because that's exactly how it was for us. My dad may have been working at Humphreys, Uncle Rob's gang may have been at Brooklands Station and my Uncle Dave's gang may have been at Glenora, and in the evenings they'd all come together. All the vans would come in from all directions and park up outside the pub; I'd be in one van with someone's kids, and Robbie and that would be in the next van, and none of us were allowed to get out of them . . . and then our Uncle Dave's oldest son Luke would be in the van down there with some kids and we'd all be out the windows talking to each other, and the workers would bring us chocolates, chippies and drinks, whatever our dads sent out – or some of them: we were favourites to some of the workers, so they used to make sure we were looked after. But that was cool; really awesome stuff. They weren't there all night, obviously, cos they had work the next day. We were only there because we wouldn't let them go without us. And then when we did stay behind . . . we would always wake up in the morning with the crinkle of chippies and chocolates under our pillows, so we knew our dads and that would bring back stuff for us and put it under our pillows.'

Hayden also remembers a comment his dad made after someone had come into the shed and tried to take him on:

'"They've got to learn to come out for a hard night as well, and then be able to do it again tomorrow – and the next day, and the next day." Cos it's true: they would work hard, but they would party hard, too. Yeah, the rattling of crate bottles or flagons was a familiar sound. You knew the day was ended when you heard the rattling of bottles . . .'

Before the boys could shear, there was an apprenticeship, much like in Johnny's day. Rousie, sheepo, presser, crutcher were all jobs taken on by the boys, as well as jumping at any opportunity to finish the last side. Robbie thinks he was his dad's favourite, as he went straight to pressing, rather than having to rousie first. 'I got to go straight on a press, which was cool; I loved it. I was probably presser and sheepo everywhere I went . . .'

The boys' fathers, uncles and other shearers were all there to guide them in the transition. The first step to becoming a shearer for Hayden has stuck with him:

'I don't think many people would have seen this, but I still clearly remember my dad – when I wanted to be a shearer, he cut me some moccasins out of the old sacks. So, I put my foot over it . . . and he would cut around my foot with the handpiece to make the moccasin, and I remember feeling it under my feet.'

For Patrick, the first step was through the help of one of the other shearers.

'Crutching, too, was good. Say you had a shearer that you liked: he'd give you his handpiece – "Nah, I don't want it now; here, you have it" – to start off your new shearing gear – "Oh yeah!" – and after that we'll go crutching with it, and, "Dad, I want to shear" – "Okay."'

Patrick's uncles stepped in when they saw him struggling.

'I had an uncle, he was pretty good with gear; he had a look at my combs one day, wasn't grounded properly, so he says, "You must be doing it hard, boy" – "Oh, I feel alright" – so I learned how to grind off him. Then Uncle Pat, he was putting the numbers out and he goes, "Don't rush, boy, don't rush – y'know you'll get there one day. Just take your time, it'll come."'

Robbie, with pride in his voice, recalls gaining his first stand:

'I could shear a hundred . . . and I was pressing; I could do a hundred and fifty, but I was never allowed a stand until I was sixteen, seventeen. Then he finally gave me my stand. Well, actually, he was a shearer down, so he gave me the chance to prove myself. I done two hundred that day, and he was like, "Okay: if you do it again tomorrow you can stay on the stand." I thought, yeah, yeah, and I just managed to get it – but the next day, man I was sore. And then he was like, "Get up" the next day, and, "You do it again," and I done it three days in a row, and, "Good on ya, you're away now." So I stayed on the stand ever since then. I did three hundred, four hundred, five hundred – every year I'd go up a hundred. It was quite cool.'

Robert's girls, Melodie and Tia, also share memories of working with the family. School holidays were spent in the sheds; both girls attended St Joseph's Māori Girls' College in Napier, and it is fair to say neither girl was much enamoured of school. As soon as they could, they were in the sheds. Melodie worked as a woolhandler first, then after she was married with small children, she became a cook on stay-outs. Good food is an incentive for those not so keen on staying out to put their hand up. Melodie is proud of the fact that 'a lot of people love our stay-outs – they're fighting to go on these stay-outs . . .' Being in the cookhouse meant she could have her children with her. She still cooks today, and she has a well-honed routine that ensures the gang are well fed. As she jokingly elaborates, 'When you're on to it like me . . . yeah, I could have tea done at breakfast time, and when they come over, make out I'm working, then go back on my bed playing on my phone.' All joking aside, providing three meals and two smokos a day requires planning and the job keeps her busy. She is quick to say how much she loves it and enjoys being able to help her brothers' business.

Melodie's menu is a little more elaborate than Ivan's.

'Yep, the normal: what I like to cook. I do ask them what they would like me to cook for them, too, and for breakfast it'll be chops and gravy, scrambled eggs, spaghetti, toast, and then I've got a pot of porridge and cream and brown sugar. I've got other cereals out if they want – Weet-Bix or cornflakes – all that's set out, but that never gets touched; they'd rather go for the hot food, but it's there just in case. Lunchtime, some like to have cold cereal because they're thirsty and hungry. I do breakfast, two smokos morning and afternoon. Morning smoko is mainly sandwiches – egg, tomato, luncheon – and some fruit. Lunch will be a variety, but they like roast with three different kinds of salads, and bread; and then afternoon smoko's usually something sweet – like, I could make a cake or biscuits with fruit. Tea is usually chow mein and dessert, and that could be a chocolate sponge pudding with custard, or if it's hot they'll just have fruit salad and cream and jelly. And that's my day.'

Tia is emphatic about where she belongs. 'I was born to be in the shed, I think.' She remembers the dynamic once her mum had introduced her to the shed at five years old:

'We used to run around with all the woolhandlers cos they . . . used to take us under their wing. I tell you, Mum had it real easy because they used to take us everywhere. We never actually had a home cos they used to keep us with them and feed us, shower us, and then I used to sit there and watch them rousie, too, cos it was awesome how they used to throw the fleeces. They were a team – a lot of them used to do a lot together. Uncle Pat was the social one, and every time we come back into town he used to have the drinks at his place and everyone gathered there. Uncle Dave used to turn up – used to be the cool one, just turn up with his car cos there was a lot of rousies, single women. And my dad used to be the one that everyone used to jump on cos he was like the big teddy bear. They all had their roles, and they all got on well with all the workers, so I was lucky to be brought up in that.'

Today Tia is a head rousie and likes to ensure the shed flows well. When necessary she can fill in as a presser as well, if Robbie is down one. The only thing she can't do in the shed is shear. She's unsure why, but her dad was not keen to see her pick up a handpiece.

The girls brought friends down from school in the holidays to work in the sheds; one was to become Hayden's wife. Hayden happened to be working in the shed that Stacy Shailer was rousieing in and jokes, 'She was after a handsome presser; that happened to be me.'

Stacy has fond memories of working with the Hape gangs:

'I was really lucky because I knew the girls before I got there. There was a sense of family. Because we were all quite young at the time, we were only sixteen, and the shearers had their wives and their children with them. Hayden's sister was our cook and she had a little boy – he was only one at the time, Cameron. The feeling around it all was very family-oriented: you'd eat together, they'd drink together . . . There was a lot of drinking, but it was really whānau-oriented, so you just become one of the family.'

Stacy also the remembers the jokes played on the naive townies when working in the sheds:

'They'd always try and get you because you were new and you were a townie. If the fleece was wet they said, "You can't press the fleece if it's wet." I remember Hayden's cousin Marama saying that "you have to hang the fleece up when you take it off the table". I remember I sort-of knew that she was joking, but – well, I probably didn't *know* she was joking, but I was looking at her warily – and she goes, "You've got to go and hang it up on the nail to dry." So, I went and hung a couple of fleeces up, and I remember Hayden's uncle coming in and going, "What the heck is that!" Then I knew she was having me on . . . I'll never get over that, but it was very funny.'

Stacy laughs as she recalls being tricked into skirting a maggoty fleece:

'I can remember trying to skirt maggots as well – if you don't know, this is what
happens to you . . . they'll try and make you fleece it, but really you should just
push it down the hole with the broom. And of course, we do what we are told cos
we have no idea what is right or what's wrong, so they have a bit of fun with you.'

A free-for-all ensues when Molly and her children remember the mischief
they all got up to on stay-outs. All five seated round the table jump in,
laughing as the memories unfold:

'Hammering their moccasins to the floor so when they go and get into them the
next day they're stuck. I've seen some rousies – [we'd] sew the leg of their pants
up so when they put their feet in they can't put them through; even their jerseys,
so they can't put their jerseys on or hoodies. Or you tell the shearers you're going
to make their bed, but you fold the bottom sheet up halfway and when they get
in they can't put their feet down. Rousies chucked in troughs . . . crayfish in the
beds . . . brooms tied up to the roof . . . This one time I remember Robbie was
shearing; he was in his prime, going flat-tack on the sheep, and then instead of
having a lunchbreak he and another couple of young shearers – might have been
you too, Pat – running up and down the board trying to chase this possum . . . and
all these older shearers are just sitting doing their gear and these ones are hanging
off the top – "Get him, bro." We had a new shearer there and they're going, "Do they
do that all the time?" But, "Oh, this is nothing." Working all day, going hard-out . . .
at the end of the day we're out in the paddock playing touch rugby.'

There was eeling, too. Robbie remembered that he stayed up all night
eeling once after a full nine hours of shearing. With no watch, he had
no idea what the time was, and got back just in time for the next day's
shearing. He may have had to catch up on his sleep during his breaks.

Robert eventually took over the sole running of the business as it
could not support the three brothers. Robert's failing health meant his
boys, Robbie and Pat, gradually took over its day-to-day running. Nursed
from home for eight years, Robert liked to keep tabs on all that was going
on and would encourage staff to visit him. Two years before his death in
2015, Robbie and Pat took over the business. Robbie now oversees the day-
to-day logistics and shears only when needed, which is generally during

the main shear. Pat is a ganger, heading his own gang, responsible for driving his team to work and ensuring the shed is running well. He still manages to churn through around 80,000 sheep a year, 20,000 of them while working in Perth for three months.

It hasn't all been plain sailing. There have been ups and downs along the way, but the family looks positively to the future as the boys aspire to grow the business back to the size that it was in their grandfather's day. A sixth generation of Hapes is now entering the sheds, ensuring that what was started over a century ago will continue. The family recognises that shearing is truly in the blood, and it is an industry they love. The significance of their shearing heritage is not lost on them. Pat reflects that he shears on some of the same stands that his dad shore on, and now his son is watching him: 'It's goosebump stuff.' Melodie offers, 'You feel like they're watching you, the older people . . . you can actually feel the older people and you know they used to walk the same walk we're walking, so that's what you get every time you walk into any of those sheds.' Molly adds, 'It's like a cloak coming over you and keeping you warm.'

Such warmth is echoed in Richard (Dick) Winiata's memories of his work in the sheds around the East Coast, despite his often brutal honesty about the roughness of the life. It is the close friendships that Richard talks of particularly warmly, as well as the shearing humour, which he recognises not everyone understands.

Another sheep shorn by
Daniel Hodgson of Mackintosh
Shearing, Meringa Station,
Taumarunui.

The lanolin from a sheep's wool
builds up on the shearer's gear
and needs to be cleaned off
regularly.

Richard Winiata –
Mahana

*'They used to say that my father, Tuki Anderson, used
to make up a bouncy swing thing and put it by his stand
cos they reckoned if I didn't see my father I'd play up.
And it had to be that close that I can touch him. So, I'm
right on the board behind him while he's shearing, yeah.'*

When Tuki asked his son in 1978 what he wanted to do when he left
school, Richard, or Dick to his mates, told his father he wanted to be
a musician. He was promptly told, 'That's not a real job, boy.' The options
were the army or shearing, and Tuki made the choice for the seventeen-
year-old, 'cos you can't take orders – you're coming out shearing'. Richard
chuckles at the memory; 'And that was it.' So the boards that Richard once
bounced on became his workplace.

Richard started working on his father's shearing run around the Gisborne area. The run had been a part of Tuki's father-in-law's, and they shore on rehab blocks (blocks of land balloted to men returning from war) that stocked around 1200 ewes plus replacement hoggets. Tuki had just the one gang and the dates for the annual run were set each year. Richard recalls there could be a bit of poaching of others' sheds:

'There was a bit of that going on, poaching; you'd call it cutting the other one's throat because you were taking someone's livelihood away from them at a cheaper rate. And when they poached, and they got the shed, maybe the first year was cheap, but the second year made up for the loss on the first year. That's just how they did things. But you knew who these people were and back in the day, when my father was around, there was no gentlemen's "I'll go and have a few words and we'll try and work it out" – it was, "I've had enough to drink, I'm going to punch his head off now." And it was a fight, and the best man won – like the cowboy days, you sorted it out. Now they'll just take you for assault and that sort of thing, but it was just part of the era, it was quite common.'

As Richard reminisces, he paints a picture of shearing-gang life on the East Coast:

'I don't know about other areas, but in Gisborne it was family-oriented – it was like the movie *Mahana*. You got the grandfather, the sons, the daughters, the nieces, the nephews, all working for one, and not till one of the opposition marries another one of the other team, then you get intermarriage, it starts to split and kids go either way. Yeah, so it was aunties and uncles all working together. Work was tough. No matter about back then or today, shearing is tough, but the environment they're working in, they tried to block out the toughness – "We're family, we'll stick together, we should be right," you know . . . They had their own judicial way of working things out, and if they did have a fight they were back on the board the next morning, five o'clock, working together – black eyes, a bit bruised – and carry on. They didn't take it too personal – it was just letting off steam. Sometimes that's all they did when they went to the pub: look for a fight, let some steam off.'

Life in larger shearing gangs could get a bit rough, too:

'Being family, there wasn't much tension. It wasn't till I got to the bigger sheds, I could see the tension – not so much the shearers, but the girls, the wives, getting jealous. They spotted the husband looking at the arse of another woman and that sort of thing, then they go to the pub and it all comes out when a bit of firewater goes in their throat; next minute you see them with black eyes . . . Sometimes, it's authority: ego got the better of them. "I've been in the shed longer, you shouldn't be my boss."'

Richard explains why the workforce in the area was mainly Māori:

'Back in the old days the landowner was Pākehā; you had the workforce that was Māori, and that's just what it was . . . until there was integration through marriages, and then you got the mixed blood. But when they were breaking in the land the landowner was Pākehā; that's all it was.'

He goes on to say that interactions were not always warm. 'In my grandfather's time it was that strict that the cocky was like Queen Elizabeth the First.' Richard tells another story, which he heard from contractor Red Fleming, about 'a fellow up here, old Jim':

'Cook used to call Red "Whero [red in Māori]". "Whero, the bloody meat's gone off in the meat safe." Red went round to the meat safe and he said, "Jeez, bloody flies in there," so he went to see Jim. "Jim, you put the meat in the meat safe and the flies have got to it." And Jim said, "Yes, ever since the boys have used it as a bitch house it hasn't been the same."'

Richard's own memories of some of the shearers' accommodation the family stayed in show that life could be basic at best:

'We'd go the night before or the afternoon before, depending how far the shed is. We get the quarters set, cleaned out, they were a mess; farmers weren't going down to clean the quarters. Sometimes the farmer leaves the window open and the possums get in, so she's hell of a mess. If the animals – rats and that – make a mess on the mattresses, well it's just turn the mattress over; the farmer ain't going to get a new mattress. Check if the water works: sometimes you may get half a dribble out of one tap, "Yeah, this one works," and, "This one's got a bit of colour in it, looks like rust," so we leave the water to run. Check the stoves; my mother's there or my wife, washing down the stoves. Sweep out the quarters, mop it out;

sometimes we used a hose, not just on the floor, but on the walls and everything . . . Set things up, the cooking gear and the kai; the presser has to go and get the meat (which is still frozen) off the farmer cos we've got to have breakfast in the morning. So, the presser or the sheepo cuts the meat up. Some places, you got the old copper water-heating where you've got to chop the kindling, light the fire and heat the water. Some places the water is like gold, so we've all got to have a bath in the same water. It gets pretty . . . you know, you can walk on it after a while. If you got a mattress, oh, that was a bonus. It wasn't the mattresses you get today; some parts were missing – kapok – big lumps.

'So that was the night before. Get to work, go and set up, start shearing, and you're still trying to figure out what gear to use. You do your first run, go back to the quarters, have breakfast, and you're sort of stunned. By then you realise, Jeez, this is scum sheep here. It's all right if the cocky is good, then it makes things a bit easier; but if he's using standover tactics, it's no good. At the end of the day you knock off and have a bath or a shower; if you're lucky you've got the shower. Some of the old fellas – I didn't when I was young, but I did when I was older – they drank. Then after a while I couldn't go to sleep, so I used to ask my father, if we were near a river, if I could borrow his spotlight and gaff; I used to go out eeling, so I could get tired to go to sleep. Well, it wasn't till later my grandfather said, "Your last run, you don't work as hard as the other ones otherwise when you finish you're still worked up and you can't come down till later," and that was it . . . so then I had to go out eeling. The old shearers that were with us, they were Tūhoe; when I got the eels, they loved that kai. I didn't eat it; I just loved getting it. That was mainly what you did; then when you cut out the shed you cleaned up . . . and then you went to the next shed quarters and did the same thing all over again.'

Richard remembers that when the shed was finished for the year, some farmers would provide beer, 'but when they came over to see us, oh, they drank our beer,' he adds with a laugh.

As to the meals, Richard recalls the lack of variety:

'Oh, it's changed a lot, yeah, yeah. Grilled chops for breakfast, roast mutton for lunch, and boil-up for tea every day . . . cabbage, kūmara . . . but the bread was all right; if you had a good cook, she knew how to bake bread, so you weren't relying on shop bread . . . you had scones or something; and at smoko time, if you had a good cook she might do pikelets or scones . . . She is the most important person in the gang: if you got a hopeless cook you got some upset people. You got a good

TOP Mangaroa Station's neat and tidy shearing quarters. As Richard Winiata remembers it, not all shearers' quarters were created equal – some were definitely not fit for purpose.

LEFT Richard Winiata stands beside his old 1949 V8 Ford Bonus, used on the shearing run by Richard and wife Tammy as well as Richard's father. It was nicknamed Henry, probably after Henry Ford. *Ruth Low collection* cook, the old workers'd bend over backwards for that cook; if the cook says, "Oh, that boy won't light my copper up," if there's a hard shearer there he'll go there and sort him out.'

Richard comments with pride that his wife Tammy was seen as a good cook, and others less so; and he shares a story that still makes him laugh:

'My mate keeps on telling me about it and I always laugh. Marty Finucane, he would say when he was shearing for Jim Godbolt back in the day, him and Derek (that was another mate I'd go and help out). He said, "Yeah bro, fuck we had this fucking cook; went back for breakfast and the whole bench was toast, buttered toast, and she had a big fuckin' pot on the element boiling away there . . . oh shit, this is good." But when they took the lid off, there was one can of baked beans sitting in the bottom. So I always remind Marty and Dorky [Derek] about it: "Can't be as bad as that woman with the baked bean can . . ."'

Humour was obviously a feature in the sheds and on stay-outs, but Richard says that it is easily misunderstood; there's a real warmth behind the sometimes crude familiarities:

'Their humour, people can't understand it: you'll either take it in and laugh, or you're insulted by the thing. But shearers, because they work together – we may not be related, but the brotherhood of being in the gang, with all our nicknames . . .
   'Shearers have a unique humour . . . they don't hide things, they say things straight up, and some of them aren't too good with their words – hence how blunt it gets after a while. It's just the humour. The friendships that you create with other people – not the marriages, just your mates – they last forever.'

One of the biggest changes to the industry Richard has seen is fewer stay-outs. He regrets the impact of this on the social structure of a gang:

'If there was a quarters when I started, you stayed at it. Whether it was five minutes away or half an hour away, you moved up there and you stayed up there. Now it's just as easy to hop in the van, travel five minutes up the road, do the work and come home. The mentality back then was different, and that's how you got that family environment that was so strong with the stay-out crews. Cos you didn't just live with people – you lived with their problems, with their ups and downs. It was

like they were your family; that's how it felt, even though they weren't. Now . . .
you're just so-and-so in the van – you don't really know them. But when you camp
out, they open up and they tell you something about their lives – "Oh, you're
married with kids" – "Oh, they go to so-and-so" – "Oh, my boy worked over there"
– and it progresses like that, eh. When you're just sitting in the van, you're just
worried about coming home and having a few cold ones . . . You don't really get to
know people, and tomorrow you might go out to the van, oh we've got a different
crew altogether. So you're at stage one every day. When you're at the camp-
out, tomorrow you know a little bit more than what you did yesterday about this
person; and at the end of the shed, no matter how many days you were there, your
tolerance level is a bit higher for that person.'

Richard continues to reflect on his time in the sheds; the significance of
the friendships built on stay-outs has clearly stayed with him:

'It's learning about people. Some of these friendships stay for a long time . . . which
is good. And that's why, when an old shearer dies, you get a lot of strange faces . . .
from those days. When they hear and they come and see you, come and see this
woman and man before they bury them . . . that's the respect they bring back
to you. So it doesn't end when the season's over; it's still there. It may not be as
strong, but it's still there.'

Sadly, Richard Winiata passed away suddenly on 24 August 2017.

While Richard reflected on the bonds that shearers make as they work
and live together, Brian Kerr recalls some of those he has encountered over
his many years in the profession – and in doing so, he reveals an astute
understanding of how to interact with others. The strong friendship
between Brian and Sam Jefferis, a farmer he contracts for, is shown as
together they tell a story of a mate's mishap with booze and paint.

Brian Kerr – 'We were all family'

Sam: 'Frank Storey – "Peanut" was his name – he was one of those tug-of-war guys. For a chap he wasn't tall, but he had muscles on his arms like this. You'd finish shearing and he'd have a bottle of beer and he'd be doing one-hand press-ups, and climbing round the rafters chasing possums. Talk about health and safety . . . you couldn't stop him. One day we had this rugby reunion in Te Kauwhata, and Peanut and Bill and all them had come . . . At night-time Peanut went home, and Brian and another guy were heading out for shearing the next morning, and they get by the Waiterimu School and, "Oh, there's Peanut's car," and it was a Valiant and it was on its side.'

Brian: 'And I said, "I better get out and see whether he's alright or whether he's in there or not." So I get up the back and look inside, and next minute this white thing moves and then these two eyes lit up, and it was Frank. And what had actually happened the day before – he was alright, thank goodness – but he had a flash suit on, he had this white paint for painting someone's fence and it had busted open when he tipped over. It had gone all over him and, instead of getting out, he'd just slept the night in this car. So we hauled him out and we rang up Bill, because Bill wasn't at work – he'd been at the rugby. He came along and picked him up and he was freezing . . . Man, it was so funny . . .'

There is laughter, so much laughter, when shearing contractor Brian Kerr and farmer Sam Jefferis get together. Brian has spent decades shearing for the Jefferis family at their Waikato and Thames properties, the last twenty-seven-plus years as the contractor. As the men sit together at the dining-room table at Sam and wife Nireen's home, the stories flow.

For years, Bill Fulton had been the Jefferis family's shearing contractor, and when he retired and went into the trucking business his son took over for a while. Then Brian, who had worked for Bill for years, took up the run. Sam jokes, 'And Brian rung up and said, "I'm taking over the gang," so we all raced round and gave him a big cuddle and carried on . . .'

Sam explains their dynamic:

'Basically, when we're shearing we're in serious mode – we don't play around. At about lunchtime, then we start joking and telling stories and laughing; but when we're shearing we're serious. We can talk while we're shearing, but we are in shearing mode and that's the thing we get done and finished. When the job's done, then it's time to party, as they say. We provide our shearers with a couple of beers each and we sit down and have a talk and laugh. It keeps people together, because some farms or stations, when they're contract, it's them and us; around here [with our] smaller sheds and local shearers it's not: it's us. And that's what we enjoy about it – that we are family. When Brian comes, he brings his daughter, and it won't be long till his grandson'll be coming down with us, too, cos his daughter and my daughter went to school together. That's one of the enjoyable parts about . . . the local gang and the smaller gangs that you don't get in the bigger gangs.'

Born in Te Awamutu in 1950 and raised in Pirongia, Brian knew what he wanted to be: 'I always had it in me that I wanted to be a shearer – I don't know why, but I walked in the shed when I was a little fella . . . and I knew that I wasn't ever going to be milking cows.' After finishing school, Brian worked as a shepherd, and didn't actually take up the handpiece seriously until he was around twenty. Although, if he'd listened to Godfrey Bowen when attending a shearing school Brian might never have stuck with it: '"You're wasting your time," he said, "you'll never make a shearer," because in those days you had to be big and everything else. I don't know whether it was a good thing or a bad thing, but I was going to be a shearer anyhow.'

As Brian recounts his shearing experiences, what is revealed is the importance of relationships – whether it be the friendship with a farmer he works for, the times spent laughing with your shearing mates over a beer, dealing with the cook or taking on the role of ganger. He's worked alongside some of the greats – Ken Pike, Martin Ngataki and Eddy Reidy – as well as working for shearing legends Bing Macdonald (an absolute gentleman, according to Brian) and Colin Bosher, who was a showman. He even recalls Colin's trick of being able to regurgitate several varieties of drink in the order that he'd drunk them.

And it seems that in the old days a fair amount of beer was drunk. Colin, he remembers, had a big shed at Pūkawa with more than 60,000 sheep to be shorn in an eight-stand shed, and Bing had a large shed of 58,000 sheep at Kuratau. They would join forces to shear the sheds just before Christmas.

'Bing and Colin used to put on this huge party at Kuratau Station because they had the best quarters, so if you flaked out you had somewhere to go. It went on for a couple of days, mind you, and some people didn't leave for a week. But they put all the booze on for ya because there were so many shearers and rousies; and the whole of Tokaanu, and Pūkawa, and some from Taumarunui, all the shearers would be there. Oh, hundreds. At night it was just one big booze-up . . . and then everyone would go home for Christmas.'

Another highlight Brian remembers was getting among the shearing fraternity at the Golden Shears:

'We used to come up from the South Island, and we'd get there the Thursday night and have Friday and then watch the finals on Saturday. That was the thing – shearers in the pubs, it was unbelievable, you're all together. And when the finals are going in this blimmin' hall . . . there's cheering and shouting. Yeah.'

Before he married, Brian spent a lot of time on stay-outs, and he reflects on the close bonds that were formed:

'Living out in a shearing gang is a different life. You become a family, there's no doubt about it. When you get married and you're going home every night, you're not living in the quarters, so there's a big difference. You get very close to people because you're there for quite a few months. Like, with Bing I used to be there all year because I used to fence at Kuratau and Pūkawa. I would stay on with Bing, me and my mates, whoever wanted to, and we'd fence during the winter and then crutch and that. We were living together all year for quite a few years, so we were all family. And I seen his kids grow up and come down; some of them are still shearing today. They do more travelling from here and there instead of living in the quarters. Having that life . . . you get out of bed, and the sheep are baaaing all night; and our cook, "Come on." There's a lot of sheds that still do it, but not as much as in our day. But that was the life. If you were going to work in a shearing gang before you got married, it was a pretty good thing. Married, to me, wasn't really the thing; it wouldn't cut it, cos it was pretty rough – parties, you know.'

Brian laughs as he thinks back on those times. For years he was a ganger for Bing.

'The ganger is usually a person that the contractor knows well. He doesn't always have to be the fastest person – I was never the fastest in our shearing gang, not out in the big gangs – and yet I ganged for Bing for years. It's just someone they can trust and know that you're going to keep the tallies, keep the money right and be honest. And if something does go a bit wrong, stop it – you know, don't encourage it.'

The ganger is responsible for getting everyone moving in the morning when on stay-out.

'So I used to wake up at four o'clock in the morning, go and put the jug on, and

TOP Awaiting replacement, the old woolshed may have seen better days, but it speaks to all those who have gone before. Jefferis farm, Waerenga.

MIDDLE With a cough and a splutter the old 5 hp Petter diesel engine that drives the overhead shaft kicks into life.

BOTTOM While most farmers have changed to modern shearing plants, in pockets of the country the old overhead driveshafts still whirr. Brian Kerr and a small crew work in the Jefferis family's old woolshed in Waerenga.

then you got to be gentle with people. You know, like. I hate getting woken up with this yelling and banging and carrying on; you just go and knock on the door – "Twenty past four" – and sometimes you might hear, "Oh yeah." And you go back again – "Are you getting up?" – and that's it, and time they get into the cookhouse, there's toast, there's a cup of tea, and then after that you waddle over to the shed; and at about a quarter to, set your handpiece up and then you start your day . . .'

If Brian trod carefully as ganger, there was an instance where he was left completely in the dark as to how he had upset the cook – something a shearer should never do.

'Over the years, what I have found is, the most important person in a gang is the cook. I've had a few bad cooks . . . I don't know what I did but I upset this old cook. She knitted everybody a hat and she never knitted me one, and she said something to me, she said: "I'm going to poison you, you —." Some terrible language come out her mouth, and I was even too scared to eat my blimmin' porridge cos I thought she was really going to poison me.'

That blip aside, generally speaking Brian knew how to get into the cook's good graces; and again with much merriment he gives away his tactics:

'What I used to do is say to them, "If you do my washing I'll buy you smokes," or, "I'll buy a carton of smokes for you a week." And I tell you what, boy, they used to lap it up because they had nothing else to do. They could have these old coal ranges and they'd just stick the leg of mutton in there – piece of cake. So that was the first thing. The next thing is, at night, if you're having a cut-out or you're wanting to go to the pub for a quick beer before tea, you never really drink too much before you're having a big feed. So you say to the cook, "I'll eat my tea later" – "Oh yeah, I'll leave it in the oven for you, it's all right, away you go." So you can whip away down to the pub, and all these others, they'd have to eat their tea, you see – "I'll do my dishes when I get back." So you got to keep right with the cook, otherwise if you went to the pub and didn't have a feed and she decided to lock the kitchen, it's tough bickie, mate, cos you're not going to get anything to eat. So that's the first thing: keep the cook right. It's most important, because they can make life pretty miserable.'

With decades of shearing under his belt, Brian has witnessed plenty of change – whether it's the quantity of sheep, the breeds, the clothes that shearers wear or the gear they use. He talks about it all with oodles of humour and laughter. Even now in his late sixties he still listens and watches what other shearers are doing and admits to having 'changed a few of my blows' recently. He acknowledges there is always something new to learn. When asked what keeps him shearing he answers simply, 'It's just the job, really: I love it.' Again, there is laughter as he says, 'I've been shearing for forty-eight years, so I'll see how far I can go before I retire. The grandkids might have to carry the oxygen mask around with me with a couple of hoses bunged into me – "Come on, Grandad."'

As Brian has confirmed, the heart of any stay-out gang is a good cook. If there's not good tucker, then invariably there is an unhappy gang. So, given their high status, it seems appropriate to include here the story of one of the country's top shearing cooks.

Ann Robinson –
'The David Fagan
of cooking'

For over twenty-five years, Ann Robinson – or Pappy, as she is known – has been feeding the shearers at Mackintosh Shearing in Taumarunui. Her name among the shearing fraternity is synonymous with good food. This good-natured, diminutive cook not only feeds over 100 people daily during main shear at the Mackintosh base, but she also acts as camp mother, counsellor, confidante and friend to the dozens of permanent and itinerant staff who work for one of the biggest shearing contractors in the North Island.

From a large Māori family of Ngāi Tahu descent, Pappy grew up on Banks Peninsula and has warm memories of time spent at Ōnuku marae. Her training in cooking goes back to her childhood, when her mother taught her all the basics – although judging from her account it's fair to say that, as in any mother–daughter relationship, she may not have always been a good listener. 'And the way I cook some of the things I cook today is just the way Mum used to do it. I remember she told me off once how I made the cheese sauce. She said, "You sauté." She was making oyster soup, and you sauté the onion and butter, and I just chucked the whole lot in; I didn't sauté the onion . . . I always remember Mum, so I sauté the onion.' Pappy giggles, remembering her waywardness.

Inspiration for baking came from her maternal grandmother, Meri Tainui (Bunker).

'She was always feeding people; she fed the community with what she had. She always had a stack of tins, cake tins on the bench, with a jar of cordial. And every tin had something different in it. That's where I get baking from. She remembered all her recipes . . . she'd do it once and then just remember it. When she became blind, she used to feel along the bench, and she still used to bake . . .'

Pappy's early attempts to mimic her grandmother, however, were not necessarily noteworthy. She would sneakily bake while her mother was off on the bus to Akaroa for groceries. Things were fine if the baking was successful; but, Pappy confessed, 'if it didn't turn out I'd throw it away down the paddock, and we so couldn't afford to waste food cos there were so many kids. But I was a bit naughty. And I learned to bake that way.'

Pappy's life changed with a shift to Timaru with her mother. After finishing school, she found work in several local hotels, cleaning, cooking and waitressing – and armed with these skills, when it was suggested she take on cooking for shearers, there were no shocks. Her first taste was up at Cheviot with Don Davies, someone Pappy remembers with much affection. Then she went on to work at Grays Hill Station with Toby McCarthy. Certainly, reminiscing about those early days there is a sense that Pappy had the skill to take things in her stride. She was responsible for the shopping and worked without a menu. She's quick to say, 'But I like menus now – like, when I go out on stay-out it's good to have a menu, cos then you just take what you need, and you don't come home with anything.' Of course, mutton was the order of the day when she started, so she learned to be creative – she would make schnitzel, colonial goose from a boned-out forequarter, crumbed chops, braised chops, mutton steaks, beef olives (or really mutton olives), and she'd do marinades and stir-fries.

From working with Toby, Pappy moved on to work with Charlie Brophy, and cooking at Otematata Station became a feature of her life for twenty-five years – first for Charlie, and later Adrian Cox. She slipped into the cycle of working July through to October at Otematata, Morven Hills and Moutere – the three sheds taking up the whole season. Her young daughter Rachel went along with her until she reached school age. Pappy giggles at the memory of her daughter: 'She used to go and sit up on the rafters in the shed and sing flat as, cos at Otematata Station back then Joe Cameron didn't like music in the shed; the only person that was allowed music was the presser . . .' Joe was always concerned that with loud music playing you wouldn't hear if anybody was hurt. Obviously, Rachel caterwauling in the rafters was no distraction.

It was at Otematata that Reg Benjamin was to experience Pappy's cooking. He enthuses that a good cook is 'just like a gun shearer':

'The famous ones, you can mention their name anywhere amongst shearers. Oh yeah, Pappy Robinson comes to mind, and she was a legendary cook . . . she is like the David Fagan of cooking, she's a legend, she's grown a lot of shearers.

'That was one of the highlights of going to Otematata. Someone would say, "What are the sheep like?" – "They're all right, but Pappy's there cooking." I'd say, "We heard Pappy's coming back," and I used to ring up the contractor, Adrian Cox, at the beginning of the season and say, "All good for Otematata?" And he'd say, "Yeah, yeah," and I'd say, "Pappy comin'?" He goes, "Yeah, yeah." "Oh, okay." . . . I'd say out of all the seven shearers I bet you every single one asked him if Pappy was going to be there. And the girls, you know, they all loved her.'

Cath Hadley, a shed worker with Adrian Cox and, later on, office worker for Ewen Mackintosh, is credited with convincing Pappy to move north. Ewen, who has joined in the interview, jokes that it is his charm that has kept her working there for so long. The banter and cheek thrown around is evidence of the close friendship and mutual respect between Pappy and Ewen. Ewen sings her praises unabashedly, and reflects on her openness: 'Everyone loves Pappy to be completely honest – great with young people.'

Farmers sing Pappy's praises, too. Catherine and John Ford of Highland Station talk fondly of her. Catherine recalls:

'I remember her coming in, when I first lived here, as the cook, and she also looked after Ewen's children when they were little . . . Lots of Māori ladies get called "Aunty" – well, that's what Pappy is to so, so many people: she's Aunty. And this shearing just gone, John goes up to the kitchen to make sure that the cook's all alright, and she's in a tizz because someone had picked up the wrong trailer, so when she opens her box to find all the utensils and the cooking gear it was empty. It's a long way to Taumarunui to get a masher and a cooking frypan and a baking dish and the beater, etc., etc., so she's on the phone to Pappy; and he walks into the shed and she says, "I'm just talking to Pappy," and Pappy yells out, "Is that John?" She goes on to speaker, and next minute Pappy tells John, "Go down to your house, John, get Catherine's stuff and bring it up to the shed." That's Pappy and that's the relationship we have with her: she knows that she can ring me and say, "Hey, my cook's having trouble, can you help?"

'Pappy's fine-tuned that cooking system so well that every one of those cooks has a Clearfile, and it has every recipe and every guidance list that you could ever think of. Even how, if the oven's blown up, what do I do – cos they're out in the middle of nowhere, sometimes, and they have things blow up and you can't get an oven fixed immediately . . .'

Despite Pappy's caring nature, there is nothing namby-pamby about her. Ewen says the role she takes on comes naturally to her but is quick to add that 'she's also fiercely "don't fuck with me" – well, if she needs to, she'll stand her ground . . . no backward steps.' As Ewen talks, he looks at Pappy and laughs, 'I think you're all forward, to be honest; it's the old Akaroa girl coming out in her.'

Pappy might be a great cook, but she is also an integral member of the team at Mackintosh Shearing's Taumarunui base. When the team gathered and shared some of their stories, what was revealed was a team working towards providing a safe, supportive and caring living environment, while facing some of the difficult challenges within the shearing community.

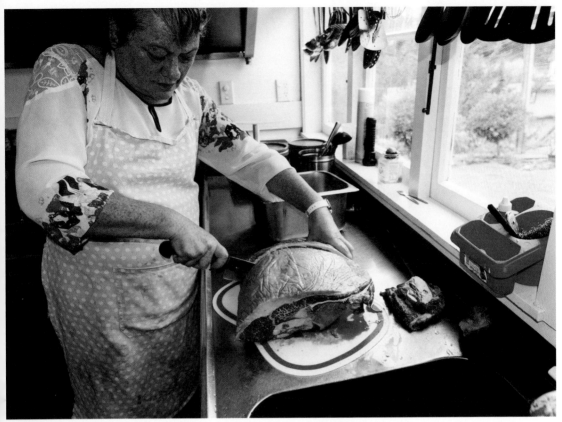

TOP Pappy works hard to produce good-quality food for the workers at Mackintosh Shearing.

LEFT While there's more often than not a smile on Pappy's face, she'll take no nonsense when push comes to shove.

Mackintosh Shearing – 'Mackland'

Ewen Mackintosh, founder and current health and safety manager of Mackintosh Shearing, is sitting at one of the long dining-room tables at Mackintosh Shearing's Taumarunui base. Alongside him are Pappy Robinson, base cook; Charlie Burton, operations manager; and Holly Woolston, one of Mackintosh's woolhandlers and trainee cooks. There is banter and laughter, particularly as Pappy, Charlie and Ewen chuck off at each other. They are relaxed and comfortable in each other's company, which comes with years of working together. Holly is happy to sit, listen and learn. This Taumarunui base has been part of Holly's life for many years. Her mother and uncles worked here, and Pappy is quick to remind her that she 'was probably two when she first came here, and she got all the baking stuff out in the storeroom and was measuring things . . .' This shared history between them all creates a warmth, and there is a sense of genuine care and mutual respect.

In 1990, when Ewen took over Alan Donaldson's run, he did not fit the bill of a typical shearing contractor. Charlie explains: 'I can remember when he bought the run and I was only young, 1990, young as, and Alan Donaldson brought him into the shed. Alan was everybody's idol – we all wanted to be Alan Donaldson when we grew up cos he was the man. And he brought this fellow into our shed at home on the farm to meet Dad and to say he's taking over the run, and we were all looking at him, me and my brother: nah, nah, this can't be right.' When asked what provoked such a response, Charlie, somewhat incredulously, says, 'Have a look at him.' Ewen interjects, 'No one can be as dumb as I look, eh?' and laughs, although he quickly adds, 'That's the mistake they made.' You get the impression that he has had the last laugh.

Ewen matter-of-factly states, 'I was never a top shearer, I was just a good average shearer,' but from the first time he held a handpiece, 'I knew that that's what I'd do probably the rest of my life; without sounding corny, that's just how it was – and I became a shearing contractor more through chance than design – it's just how it happened. And I would've just kept shearing, hopefully would have ended up with a farm as well.

'I bought Alan's contract shearing business, but there were about 550,000 sheep on it, no quarters, 'bout four or five old junk cars and one van, an old Beddie.' The 'Beddie' was an old yellow Bedford van. Ewen had ambitions to grow the business; he did not see the small business model working for him. 'I did cold turkey, just canvassed. You actually become quite unpopular doing that, but it didn't really worry me back then. It probably worries me a bit more now . . . I think I'm a bit more diplomatic about things.'

Mackintosh Shearing currently shears approximately 1.3 to 1.5 million sheep a year and runs from two bases, Taumarunui and Taihape. During main shear, from the beginning of December to the end of March, there are 130 to 150 staff working out of them. There is a cook for each base and an operational manager. The business model is slightly different from most shearing contractors, as Ewen has brought partners in, Ben and Sue Gower. Ewen says Sue is the intellect behind the business and Charlie and Ewen are the brawn. Charlie and Louise 'Buffy' Burton have become partners in the business, and up until recently Larry and Harata Clark at Taihape were also partners. Computer spreadsheets allow them all to juggle the logistics of the business.

Having the operational managers has freed up time for Ewen to focus on health and safety, a contentious area of the industry but one that Ewen is zealous about. He is the first to admit that, when he started out, health and safety was not front and centre:

'We didn't bother in the past . . . I ashamedly say that we used to not really care whether someone went to work with a drunk driver, cos everyone did it – that's just how it was done. But now I think a good responsible contractor will have good solid health and safety . . .'

Procedural manuals, earplugs, protective eyewear for grinding, farm safety checks and daily breath-testing of drivers: these are all features now, and the leadership team continues to proactively grapple with the fraught issue of drugs. Ewen is also working to encourage change within the industry through his involvement on the executive team of the New Zealand Shearing Contractors Association.

'Someone's got to lead, though, and that's generally what happened here – we just lead it, and we've got our standards. I believe the biggest thing is to set the example yourself, and if you're doing things pretty right, and Pappy feeds well, if we pay on time, pay as good as we can, house them well, cart them well, do all that – then we've got the right to go back and say, "Whoa, whoa, whoa, whoa."'

Ewen remembers when he had his epiphany. It was station manager Bill Pullen who altered his thinking.

'When he was managing Otiwhiti Station down Hunterville, I turned up there in the dark, first time I was there, and opened the door into his shed – the side door – I looked to go into the shed. It was gleaming and spotless, and he had obviously spent an afternoon polishing that shed out for us. That was probably one of the most important lessons that I've had in my life: that I had to do a good job. He'd set the standard and that's what we try and do here. I still remember it, and Charlie sets the standard, and he's got the right to go back if they aren't doing their part. That's how we see it.'

As to ensuring his staff are well fed, Ewen acknowledges the influence of another:

'I personally believe a contractor in the South Island changed it. Peter Lyon, he started feeding people really well and set the example, and generally most good contractors feed well now. I've always thought the difference in money between feeding someone poorly and well isn't actually that much, and you make it up very quickly through production . . . feed them well, and they work well and want to be there.'

Charlie chips in, 'Can't run a V8 on a two-stroke.' Ewen agrees: 'Yeah, and that's about it, too: it was a false economy.'

Another area the team grapples with is the drinking culture, as Ewen explains:

'We've tried to change the culture here, and to a degree I think we've done pretty well at it. You've constantly got to keep at it, but more and more people are starting to realise that they can't make a life out of doin' that. One of the things that Charlie brought in: he brought in breath-testing every morning for every driver, and their jobs became more important to them than their booze. When we started drug testing a bit, too, their jobs were more important to them . . . I think people sometimes are just looking for a reason to change, and with the boozing and the drugs you give them that reason; I think they're actually quite relieved at times. I don't drink, and I've always had shearers here that felt comfortable not drinking because of that . . .'

What stands out about Ewen is he is willing to acknowledge where he has made mistakes, and talks of even standing up in front of staff and admitting them:

'I had a cook here a long time ago – it was probably before Pappy – and she was

TOP Dinner time at Mackintosh's Taumarunui base. At the peak of the main shear Pappy will be feeding upwards of 100 crew at a sitting.

BOTTOM Mackintosh shearers keeping the woolhandlers busy at Meringa Station, Taumarunui.

just fuckin' terrible, excuse me, but she was. I put up with it and I put up with it, and in the end she left. And I stood up and said to the guys, "That will never happen again." I had thought that I couldn't change it and that my balls weren't big enough at that stage. I should have dealt with it a long time before I did and it was just demoralising for everyone. So Pap must be quite good – she's done twenty-five years.'

This all points to a leadership style in which authenticity counts.

'I think if your staff know you do genuinely care, they'll do what needs to be done. You've got to be authentic; if you're not authentic in this job they pick it out in ten minutes, they do.'

The talk around the table covers tough issues faced by some members of their crew. There are young people coming into the industry who come from hard backgrounds and need support. The team is fostering an environment where opportunities are given to those looking for them and where support is provided when needed. Pappy quite naturally builds relationships, especially with the young girls, getting alongside them and encouraging them and building them up. Given that there is the potential to earn well, both Charlie and Pappy talk of encouraging staff to be careful with their money, offering a bit of financial guidance where needed. They're keen to see those who work at Macks build a future for themselves and not to squander their opportunity to get ahead.

When asked about his desire for authenticity and care, Ewen is quick to point to his father's 'unconscious role modelling', which has influenced the direction he has taken the business:

'My mother's very caring too, don't get me wrong; but my father was a very caring sort of person to someone less advantaged. I remember him bringing the hobos back from under the bridge . . . he used to bring them up for a feed. We also used to be a home where pregnant girls used to come and have their babies . . . apparently Dad was exceptional with them, he built their confidence up again and made them feel that they hadn't made a mistake.'

Talk and banter continue over a lunch that Pappy conjures, and it's clear that while there are expectations and standards laid out for staff at Macks, the high level of care encourages a reciprocity between employer and employee. Nothing illustrates this better than a brief story from Ewen about a young woolhandler who tore into a shearer for mistreating a sheep. 'That was Carmen Linley, and she ripped this guy up – "We don't do that in Mackland," she said.'

The shearing industry has often had a bad rap over its 'work hard, play hard' culture, and undoubtedly it can be rough; but there is so much more than that to life in a shearing gang. There is hard work in bucketloads, and there is a lot of fun, mischief and mayhem to be had, with or without the addition of beer – but there is also something deeper. Pappy talks about it as a sense of belonging, and Barry Pullin refers to relationships: 'It is actually about the family, the whānau and the togetherness the shearing industry has, and I think it's all about that networking and relationships. Even in the shearing industry itself, a lot of people don't understand the value of those family and network relationships.' Those relationships extend to farmers as well. When farmers' and shearers' children have grown up together, or a contractor has seen a farmer's child grow up, then sees that child take on the running of the farm, that all adds to the colour and richness of the industry. For some, shearing will only ever be a job; but for a good portion of those in the industry it is a lifestyle.

Machine shearers compete at the 2016 Canterbury A&P Show (now known as the New Zealand Agricultural Show)

4

Their Guts and Glory

'. . . only a couple of blows
behind . . . he's round over
the last shoulder . . . she's
four or five blows ahead . . .
it's long blow time – right up
the backbone . . . the sweat's
pouring off them . . . give
them a round of applause,
ladies and gentlemen.'

The competitive nature of shed shearing naturally lends itself to competition in the sporting arena, and commentary such as that on the previous page can be heard at any one of the sixty-odd contests held around the country in a year. These events provide shearers with a public environment to pit themselves against each other and provide an opportunity for individuals to shine.

The Golden Shears in Masterton and the New Zealand Shearing Championship held in Te Kūiti have become the most prestigious events on the shearing calendar. For blade shearers the ultimate competition is the Golden Blades held at the Canterbury A&P Show. An international competition emerged in 1977 and, while it took some years to establish and grow, the World Shearing and Woolhandling Championships has now become the 'Olympics' of shearing; the 2017 event saw thirty-two countries competing before a crowd of 4000 at Invercargill's ILT Stadium.

Another way to compete is through world record tallies, which sometimes feature on the TV news, especially when a record is broken. The camera pans to the face of the shearer, sweat beading on skin, with that one drip that always reaches the tip of the nose and hangs there until gravity releases it, and the shearer's saturated singlet clinging to tired muscles, as the individual (or individuals, if it is a team record), goes hell-for-leather stripping wool off a sheep. Newspaper headlines, too, report the records as they tumble – 'Rowland Smith smashes world record', or 'King Country shearer breaks women's nine-hour record' (that's Kerri-Jo Te Huia) – and the name of yet another shearer enters the psyche of the general population.

At the end of the eight or nine hours, when success for the shearer has been assured, there is relief, as well as hugs and exhaustion as the enormity of what they have achieved sinks in. Among the supporters there is jubilation and pride. And for all the volunteers there is a sense of a job well done, the reward for months of preparation and scrupulous planning. For anyone in the industry, and particularly those individuals who have held a record, there is a deep understanding of the pain and sacrifice that went into making that record.

Young shearers have idolised those elite competitors who have gone before them, attentively watching the nuance of every step and blow, seeking to emulate them in the hope that they, too, can become one of the legends. As has happened in other sports, the professionalism of the athletes has increased almost exponentially as the science of sport and nutrition has become better understood. Record holders and world-class competition shearers are training and preparing themselves for events. The dedication and commitment now required to compete includes much less time spent in the pub and a lot more time in the gym. Matt Luxton, the personal fitness coach for Matt and Rowland Smith, talks of the spokes of a wheel that are in play when making a record attempt: 'A combination of mental strength, skill, fitness, support, teamwork, nutrition and relentless planning . . .'* The dedication of these extreme athletes brings not only recognition for themselves but also greater recognition for the sport and, by extension, the work of the shearer.

In bygone days, a shearer's idea of training was likely to be churning out more sheep in the shed and laying off the beer a day or two before competing or going for the tally. The memories of those in the next pages all tell their story of what it is to take on the 'sport' of shearing. West Coast shearer Malcolm Sadler was a team member in a six-stand record in 1965. Back then there was no meticulous planning: just a decision that the sheep looked good, and a phone call to the local JP.

---

* Matt Luxton, 'Getting fit to shear world records', *Shearing: Promoting our industry, sport and people* number 97, vol. 34, no. 2, August 2018, p. 29.

Malcolm Sadler – A six-stand record

'I have to say it was nearly the highlight of my life, although we did better tallies at the eight-stand shed – but I'd only been shearing just over two years, so it was quite an achievement…'

Seventy-nine-year-old retired shearer and builder Malcolm Sadler sits in the living room of his home tucked away up the Grey Valley in Nelson Creek recalling his shearing days. His proudest memory is of the six-stand record gained in 1965 at the Ferguson family's Waipuna Station. Unlike many shearers, a woolshed had not featured in his early life; instead, he was more likely to be found chasing cows in a cowbail as a toddler or milking cows after school on the family's Nelson Creek farm.

In fact, shearing, other than a little bit of dagging, didn't feature at all until Malcolm was twenty-three or twenty-four. He'd successfully finished a building apprenticeship with Western Construction in Greymouth and had been working for a couple of years. Peeved at a lack of promotion, he left. Realising he could earn more money shearing than building, he went shearing with his brother Neil, who had just lost his offsider, Paul Wolston.

Malcolm went straight into a Wool Board shearing course with Claude Waite over at Lincoln College. He remembers Claude being a 'brilliant man' who 'was really good at getting everybody into the job'. Though Malcolm reflects, 'I think it probably took longer to learn to shear than what it did to learn to be a carpenter, because there is a lot involved in learning to shear.'

Back then, second shear hadn't been introduced to the area.

'In the early days we only did the full-wool shearing; we shore a few hoggets in November, maybe, and then later in December we started ewes – which were mainly full wool – well, they were all full wool; they didn't do second shear. With our wet weather there was quite a lot of cotty sheep on the West Coast; they upset quite a few people that come, cos they couldn't handle the cott like we'd been brought up with.'

With one season of West Coast shearing under his belt Malcolm headed over to South Australia and found himself learning to use narrow gear. He shore solidly for two years between Australia and the West Coast; 'By the second year I was starting to get reasonably proficient.'

Malcolm explains the decision to attempt the record:

'I think it was just that we realised the sheep were good and we had a pretty good team, and then the joker Peyton had been written up how he'd broken the six-stand record and we came to the assumption that we can better that – and yeah, we certainly did.'

The year before, he remembers the gang getting a 'reasonably good tally', although 'it wasn't a world record-breaking tally' . . . 'and then this season we could see the sheep were in great order; and there was 'a pretty good line-up'. Colin Chamberlain was one of the shearers and Malcolm comments he had already done 511, 'and as far as I know it was the first official five hundred done in NZ'. Malcolm explains that Colin Chamberlain was Colin Bosher's top shearer:

'It was Colin Bosher's crew that used to come down and fill in the extra shearers. Malc [Malcolm] Love was one of Colin's shearers, and I'm not sure who the other ones were now but, yeah, Bob Winn filled in for the day and it was a pretty good effort for him – I think Bob did 376 . . .

　'The day before we spent mainly just timing ourselves, too, seeing what we could achieve. We'd see how many we were getting on the quarter, cos if you're going to do a tally you have to time yourself as well as your pace from your shearing mates. You watch the clock, and when you're getting eighty a run you've got to do ten a quarter; you shear three the first five minutes, then two, and then three, and get the catch at the end of the run if you can, and that'll give you your eighty for the run. But you get pretty good at watching the clock . . . to shear four hundred sheep in nine hours, I think, is one minute seventeen a sheep, and to shear four hundred in eight hours is one minute fourteen . . .'

Malcolm recalls with a grin that George Ferguson, the station owner, was on board, but only up to a point:

'My brother Neil suggested we finish a little early on the day before, so we'd be a little bit fresher, maybe, for the next day. George put the hammer down and said if we didn't work till time we wouldn't shear the next day. So, we had to work till time.

　'The sheep were in great nick: they weren't big and they were combing really well. We actually had Colin Chamberlain came down and Malc Love, and then there was Malcolm Farrell – he'd been ahead of me the year before. We walked into the woolshed together, six-stand shed, then we got on the board where I took a couple or three big steps and put me gear on stand number three. And he gave me an awful look, but I was a bit cheeky in those days; the ringer held the top stand, and if you were beaten for two days you had to move back down the board. So, I think we shore sheep for sheep for most of the way through the shed, but on tally day his gear wasn't going quite right and he only got 396. Malcolm Love said, "He's had to be useless if he couldn't have done another four." But the poor fellow was probably absolutely flat out – he said to me, "I tried every way to get round you."'

Malcolm explains how he wasn't clear on just how well he was doing:

'I didn't actually know what I did – I thought I was only doing a bit more than three hundred, you know. I said to the rousies and that, "How many would I do?" They said, "You'd do four hundred," and I said, "Gee, whatta you got to do to do four hundred?" They said, "Oh, you do eighty-five the first run and seventy-eights after that" – or eighty-six, I think it might be . . . In the first run I think I only got eighty-two out and I thought, oh, cos it's a two-hour run before breakfast and then you work four one-and-three-quarter-hour runs. I thought, "Man, I'm in trouble" – cos usually when you're doing a tally you really want to be on the mark first up, because if you start struggling or fighting you make it awful hard on yourself. But I had me mind set that no one was going to upset me and I wasn't going to rush; I was going to keep going nice and steady. And I got eighty out the second run – and, yep, I'm on it. When I hit the four hundred I said to the shed hands, who'd been egging me on all day – they drunk lemonade – I'd buy them a bottle of lemonade for every one I shore after four hundred. I think I only got to 411, but I think I bought them a dozen each. It was one of those days when the sky was purple and it could have hosed down any time – it was a real muggy hot day, which was probably really good for shearing cos the sheep were warm – but the rain had to stay off till after three o'clock for us to have enough under cover to finish the day. So it could've all gone . . . and there was only one mob on the farm: the main ewes.

'I don't think they put any extra rousies on, but we had two Māori girls from the North Island, they were absolutely marvellous: Minnie Johnson and Lucie Hauraki. They were just phenomenal: they shifted the wool away and laughed and giggled and egged you on all day . . . You're being a gang working together, and we lived at the quarters, and I think when you all lived together and went to the pub together, and we played together, you did build up associations in the crew – you really got to know your shed hands and your shearers. I think it's a better atmosphere than if you just travel to the shed every day . . .

'At lunchtime we predicted we'd shear two and a half thousand. I think Neil talked to the newspaper – it was Neil's crew – and that was what we looked like we were heading for. But we actually ended up twelve off, which wasn't a real big blow, really. But to boost the record from 2200 to 2488 was an awful big jump, and it was a really good tally because it was eight years before it was lifted again. In the 1960s and '70s records were falling over pretty quick . . . with sheep numbers up and the shearers getting more proficient, and one of the amazing things I was telling Ken Ferguson yesterday – Malcolm Love, you know that year we did the tally, was going up to George Ferguson's and drinking a bottle of Black & White whisky with George every night. We must've been there twelve days because when

TOP The old woolshed where Malcolm Sadler's six-stand record was made no longer stands, but the board bearing witness to the event has been transferred to the new shed.

BOTTOM Waipuna Station woolshed.

he [Malcolm] finished the shed he had twelve black and white horses pinned down to the lapel of his jacket, and that's got to be a bit of a constitution – and he still shore 428, yeah.'

Of course, after making the record there had to be a celebration.

'I think we got home from the pub about two o'clock. We did have a bit of a bash, but we weren't very good the next day . . .

'I think someone said, you had them too well crutched, but we did have to take quite a big crutch off them because they were quite dirty. I think with records you're only supposed to take as much as you can cover with your hand, but some of them, well, we had to even cut round the legs to cut the dags off – perhaps we had prepared them a little bit more, but it still was a good effort. But they were quite daggy, and sheep that have got fresh dags on them usually do shear much better than stuff with dry dags on them; it means they're picking up and that the wool's in good nick to shear.'

For decades Malcolm has been an identity in the Grey Valley shearing scene. He took over his brother Neil's contracting run and ran it for years. With fewer sheep to shear on the coast, his downtime was spent building houses in the district. He believes he was the first West Coaster to put out a 400 and remembers that for 'the next eight or ten years I had people coming determined to beat me – it sorta sets the pace and sets the tone a little'. The eight-stand record and a tally he achieved with Mike Boyd are other moments he remembers with pride:

'In later years me and Mike Boyd did a four hundred at Waipuna on the Monday, and we didn't hit four hundred again during the week; but our worst tally for the whole of the week was eighty-eight, and we had five cut-outs. Other than that we were shearing ninety-five or ninety-six every run. Mike reckoned from the Saturday when we were doing lambs till the Saturday night when we finished the rams, we had shorn two and a half thousand sheep for the week – it's not a bad tally.

'Sometimes when you do a big tally you finish up and say, "I don't know why I did that" – but I suspect it's like climbing a mountain: you set your mind to it. I think that a big part of the focus is to be able to set your mind and know exactly what you have to do and how to do it . . . When you do a tally it's a mental block, thinking, "Can I do it? Can't I do it?" But once you've done it and you get rid of that mental block, a whole new world wakes up for the next bloody tally.'

Huge satisfaction can come from competing, just as it can from being part of a world record. Being right in among the guns of shearing can be a real thrill, too. Reg Benjamin competed in his first show in Marton in February 1979 and stopped serious competition in 2000. He has shorn 1.5 million sheep over the decades, but he rates competing alongside the likes of John Fagan, Alan 'Mickey' MacDonald, his Taranaki cohorts Paul Avery and Ian Buchanan, and the great Sir David Fagan as career highlights.

Reg Benjamin – The sport where you're paid to practise

'I used to watch it on TV – there was only a wee segment, but I always knew in March they would have a little bit about the Golden Shears and they'd say who won, and they'd show a blip of the shearing and I remember thinking, "Oh yeah, that will be me one day."'

When Reg Benjamin's parents moved their family from Palmerston North in 1974 to buy the Manutahi Hotel in Taranaki, it's not likely they thought the patrons might influence their thirteen-year-old son's future career. Both shearers and freezing workers frequented the hotel and the young Reg soon had the measure of them.

'Shearers seemed more exciting to me. They were fit, they were strong, they were laughing, and they always had plenty of money – and they were characters and they looked like they were having fun. Whereas the freezing workers were all staunch and moaning about the work and they were going to get the union on to them and they were going to sort this out . . . But the shearers were a different bunch of guys; a lot of them had farms and they had money . . . they were guys that I looked up to.'

By the time Reg left Hawera High School partway through sixth form, he had already worked in shearing sheds during school holidays with Eltham contractor Don Bolstad. And after completing a week-long Wool Board shearing school in term-time, he was even more determined to launch his shearing career. He still clearly remembers the headmaster's reaction to him leaving:

'I handed in my leaving card to the headmaster and he thumbed through it and he said, "So what are you going to do?" And I said, "I'm going off to join the shearing gang." And he just stopped and he said, "Is that the best you can do?" And I said, "Well, that's what I want to do." I couldn't explain it to him cos he thought, raggy-baggy bum shearer. But I looked upon it as a sport, cos I wasn't a real sportsman at school – oh, I played rugby and I enjoyed the team stuff, but I never pursued anything with real passion. I liked shotput and discus at athletics, like anything that involved strength, and I was a good weightlifter . . . but shearing really appealed to me because it was a sport that you got paid to practise.'

This was in the days before professional sport in New Zealand, so Reg could see that shearers were earning big bucks, the prizes in the competitions were pretty good and there was prestige attached. And he didn't want to be like the headmaster. 'I didn't want to work in a dingy old office for the rest of my life dealing with wayward kids.'

A local event also caught Reg's imagination and sent him in the direction of the woolsheds. Alan Bolstad, Don's son, unofficially beat a world shearing record.

'Alan was a very, very good shearer. He was only young, but he was very good; he was talented – he and another guy unofficially beat the world record, just down the road from home. At that time, I think it belonged to Larry Lewis in Gisborne and it was five hundred and fifty-something-or-other, and they unofficially cracked it. And I thought, well, fancy that – little old Taranaki with a couple of young blokes. I thought,

TOP Reg helping a mate shear a small mob at a three-stand shed in South Canterbury. The young woolhandler keeps the floor clear for Reg.

LEFT The ubiquitous soap-powder buckets have become the gear-container of choice for many shearers; conveniently they double as stools.

man, there's something in it. To me it was like Taranaki winning the Ranfurly Shield. It was totally unofficial; to be official it would have to be ratified by judges done under the right conditions – but they weren't there to break the record, they just thought they would have a crack . . . Yeah, that caught my imagination, that side of it; to me it was like a sport, and that's why I didn't even try to explain that to my headmaster.'

Before he was anywhere near ready to compete, Reg first had to go through the shearer's typical 'apprenticeship' of rousieing, pressing, dagging, crutching and barrowing before finally getting his first stand. And with Don Bolstad as his boss it sounded no picnic:

'When I worked for him he was in his late forties. He was big, strong, of Norwegian descent. He was a real Norseman, too: he followed in the tradition of a Viking. He was proud of his Norse heritage, but he was extremely grumpy – he was fiery, he had a shock of red hair and he had a look that would send a chill down your spine. He was a tough brute of a man; he'd been a shearer all his life and his back was bent – he probably shore too long . . . Everything was done military style: if he told you something you made sure you heard it the first time cos he wasn't too happy if you asked him again.'

The memory of his first trip out to a shed still stays with Reg:

'Don Bolstad was a chain-smoker, my boss. He smoked, I'd say, from the second he woke up till he shut his eyes and went to sleep. He had an old EH Holden – well, it wasn't that old back then, it was a 64, but we're talking 1975; you didn't think it was a beat-up old bomb or anything, but it had a hard life – and cos he smoked he always had the window down; he didn't care if it was raining, he used to wear a big Swanndri bush shirt . . . He wouldn't care if it was freezing, cos he'd be smoking away all the way. I'd be sitting in the back, and you'd get pelted with embers and ashes and even the smoke . . . you didn't dare say anything. He was a big fiery man, you know . . . Thor, we used to call him – Thor. You didn't dare say anything, cos the scowl alone would just about stop your heart. You'd get picked up at four o'clock or ten to four; it would be dark. He had this saying, I only ever asked it once, "Where are we going today?" "YOU'LL FIND OUT!" Like, how dare you, you insubordinate thing, ask me that – "YOU'LL FIND OUT!" So that was the first and last time, but I saw it happen heaps of times with new people; I'd be trying to say, "noooo".'

The vehicle was packed, no seatbelts, three in the front and four in the back.

'The first day, I remember, it wasn't that far away, we turned up there and he sort of "RRRRRAAFFF!" He grumped and groused and I didn't really know what to expect, so I'd packed a bit of a bag with a change of clothes and that, and the boot was packed full; he said, "WHAT HAVE YOU GOT THAT FOR?" I said, "Oooh well, I thought I better bring some clothes." "YOU'RE ONLY GOING TO WORK, YOU'RE NOT GOING ON HOLIDAY!" "Oooh, okay." He stuffed my bag in – just as well there was nothing fragile in it, it would have smashed.'

Despite Don's less than affable nature Reg worked hard and eventually won his stand. 'I wanted to be on that stand shearing, you know, I wanted it badly. I knew I'd do it, but I had to put the effort in.' Reg carried on working for Don until 1984 and then worked for world champion shearer Roger Cox. Reg eventually bought both Roger's and Don's shearing runs and contracted for seventeen seasons.

Competitive shearing quickly became a part of Reg's programme.

'Right from when I first started working in the woolsheds – cos the guys I was working with were professionals, and they were going to competitions and doing well, too – to me, it wasn't just a job, it was a sport, so I was into it right early.'

Roger Cox from Taranaki was reigning world champion at the time, so Reg wore his black-and-yellow singlet with much provincial pride.

'The first shearing competition I competed at was in Marton and it was 1979, I think – February 1979 – and I shore in the Junior, the very first grade. I didn't win anything, it wasn't a podium finish, but I was right up there, and I was quite encouraged. I went to the Golden Shears; didn't do all that well there, but I was striving to do better. The following year I competed in the Intermediate and I got a few local ribbons. I won a couple of local competitions and then I started to travel out into the circuit, the lower North Island, which incorporated the Golden Shears: it was called the Golden Circle and you competed in about twelve or fifteen shows – Levin, Palmerston North, Marton, Rangiwāhia, Taihape, Taranaki, Stratford, Whanganui . . . Waverley was another one, Dannevirke, Carterton. And all those points went towards the final at the Golden Shears. So if you did that circuit and you got enough points you'd shear at the finals at the Golden Shears; that was

OPPOSITE Old wooden wool presses may still be in service around the country. They are becoming rarer so the days of clambering into the press and stomping down the wool, a rite of passage for so many, is drawing to an end. Farmer Stuart Jefferis fills the bales at his farm in the Waikato.

LEFT Woolhandler Lianna Rix throws a merino fleece at Mt Nicholas Station.

BELOW Young sheepo Alex Nicholson of Hape Shearing negotiates the maze of gates behind the shearing board to ensure a constant supply of sheep is kept up to the shearers, Mangaorapa Station.

always an aim of mine and I managed to do that a few times, right up to the Open. Competition shearing was a big part of my life.

'In the summertime the competition circuit started properly towards the end of January. Usually by that stage, if we'd had a good run with the weather we usually took the weekends off for the competitions. Sometimes there was work that other guys would do – a few catch-up jobs in the weekend – but generally our run in Taranaki was tailored around the competition season, which ran from January . . . basically the Golden Shears was the end of it in March. So there were six weekends where you'd be away and competing. Some were on Friday, some on Saturday and some on Sunday. Most weekends we'd do two in a row. Like, Pahīatua was on the Sunday and then the Golden Shears started on the following Wednesday. Everything culminated in the Golden Shears back then, until the New Zealand Champs started in Te Kūiti, which were always Easter and extended the competition scene a bit. So, once that happened, more competitions sprouted up between March and April; that made the competitions scene a bit bigger, and over the years it's come back the other way. Alexandra starts in October, and the Canterbury show, and all those shows fill in the gap before Christmas. These days the competitions circuit runs really from October to May.

'A lot of weekends there would be doubling up in the upper North Island or the lower or central North Island: there might be a show in each area on Saturday and on Sunday, and there would be something in the South Island. If Fagan was chasing points for a certain circuit he would go to the South Island, but he would always leave it a bit of a mystery where he was going to turn up in the North Island. We would try and second-guess him – you know, think, "He'll probably go to Rotorua," so we would all head to Taihape . . . Rotorua and the Taihape show were both on the same day in the middle of January. And we'd get to Taihape and we'd see his car parked there; we thought, "Oh no. Have we got time to carry on to Rotorua?" Competing with him, if he was there – it was like, write the cheque to him, he was so dominant in the sport.'

A competitive highlight for Reg was being the 'understudy' for the New Zealand team shearing test at the Australian Golden Shears held in Euroa, Victoria in 1981, 1982 and 1983. In 1982 he competed in the Individual Senior competition.

'I made it into the final, so in effect I was the first New Zealander to make their Golden Shears final – with narrow gear, too, so that was quite a feather in me cap because we were shearing totally under their conditions and with their gear.'

Reg is also proud of the fact that he had a world champion work for him when he was contracting:

'Paul Avery, he was only very young when he started with me and . . . he won the world championship in 2008. And I remember when he first started he was only a little weed of a bloke, but I could see he was going to be good. He was determined, and we used to call him the little Suzuki Jeep with a Chev 350 V8 heart; but at the end of the day I thought he was too small to make it to the top, and I used to joke to him, "Look, we're going to have to feed you two litres of ice cream every night." But he put his mind to it and he did build himself up a bit, he really trained hard and he reached his goals. He wanted to win the New Zealand championships, the Golden Shears and the world championship and he did all three of those, and I was very proud of him.'

Reg is not ready to hang up the handpiece yet; he is still churning out sheep, many of them now shorn in Australia and Europe. But when he reflects on his time competing, he is not disheartened that he did not see his name in lights. Rather, he carries with pride the fact that he shore in the David Fagan era:

'I was really proud to be in an era that included him because I'll be able to say to my grandkids one day – he is such a household name, he'll be a legend for years to come – my grandkids might ask, "Did you ever see that guy David Fagan?" And I'd say, "Oh well, I used to talk with him and compete with him." Like, we'd compare combs and things like that; he was such an approachable sort of guy. He was a champion, but he didn't mind letting stuff go – he was a great guy . . . and he was pretty liberal with the praise, too: if you did well he was the first one to come and pat you on the back . . . I'll always be proud to think that I competed in the same era as Sir David.'

Someone else who competed in the David Fagan era is Bart Hadfield. Like Reg, he loved competing too, but the dream of farm ownership drew him out of the circuit early. Nonetheless, Bart still loves the sport and he takes every opportunity he gets to coach young learners, encouraging them into shows. Bart and his wife Nuku reflect on what shearing has meant to them and the commitment that Bart is making in giving back to the industry that he loves.

Bart Hadfield, ready for a break
at the end of a run. Mangaroa
Station, Northern Hawkes Bay.

Bart and Nuku Hadfield – 'A nice place to be'

*'I'd say there's people that shear because they're good at
it and it's an income – it's their livelihood – but there's
also those that have a real passion for it. And Bart is the
one that has the passion, and therefore you take it even
further . . . hence teaching young people to shear. And
now he's a shearing judge, a record judge; we organise
the shearing at the Wairoa show. So you start taking it
to those other levels, because at the end of the day you
love the sport and you see that there are other ways you
can help it grow other than just being in the woolshed.'*

Nuku has Bart well pegged. Sitting at the dining-room table in their
110-year-old farmhouse on Mangaroa Station, Bart exudes energy
and enthusiasm for shearing as he and Nuku talk of their years in the
industry. As Bart's story unfolds it is easy to see why he took to shearing:
his father shore, and his maternal uncles shore as well. As a young lad
Bart would tag along with the men, keen to watch them in action as well
as take up the handpiece whenever he got the chance. He simply states,

'that was all I ever wanted to be . . . I always wanted to be a shearer'.

And so he was. By the age of twelve Bart could get his way around his father's sheep on the family's seventy-acre property at Whangaehu, south of Whanganui. By age fifteen he'd tallied his first 100. By age seventeen, in 1986, he had left school and was working with Lee Matson, a big Whanganui contractor. He did his first 200 on his first day on the job – 'I just hooked into it and went hard.' And that was the start of Bart's fourteen years of professional shearing. In those years he packed as much shearing in as he possibly could. By nineteen he had shorn his first 400. By twenty he'd done his 500. In 1997 he, along with Steven Stoney and Rodney Sutton, set the Strong Wool Lamb three-stand nine-hour world record. There was shearing in Australia and England, seasons down south in January and February with Ron Davies. When he moved to Taihape there was shearing with Paul Kelly, Rudy Lewis and Lloyd Alabaster. His best-ever tallies in nine-hour days were 525 ewes and 803 lambs.

As if that wasn't enough, Bart also managed to buy 160 acres out of Whanganui at age nineteen. By twenty-four he was more focused on a future in farming and was able to buy 1000 acres in Taihape, using the smaller property as a stepping stone. He then juggled shearing and farming. In there somewhere was the budding romance with Nuku, who worked at the ANZ bank in Stratford, and eventual marriage and children. To add to the mix, Bell Hadfield Shearing Contractors was established in 1998: Justin and Diane Bell and Nuku and Bart ran the business together until 2001.

Shearers all seem to be good at acknowledging the people who helped them develop their shearing skills, and Bart is no exception. He had attended a couple of shearing schools 'and they helped no end', but he attributes learning the most off his mate Keith Manunui. Keith, working for Lee, took Bart under his wing and taught him 'heaps'.

'Everyone can get the wool off, but there's efficient ways of doing it, and Keith had a very efficient way of doing it; he made a very tidy job and didn't cut the wool in half and was very professional about all his gear and his attitude towards the job. That rubbed off on me, hopefully.'

A piece of advice that Keith imparted always stayed with Bart:

'Keith used to say to me when I was learning to shear, "You should sweat on your brow before you sweat on your shoulders." Because you're concentrating on what you're doing. So that was a good bit of advice; I always used to work on my pattern, on style, make sure I was hitting it right; and then ten minutes into the run you're covered in sweat because you're trying to shear them as quick as you can, obviously, but as easy as you can.'

Competing was a natural extension of Bart's shed shearing.

'I'd seen them when I was at high school – Golden Shears – I'd seen all that sort of stuff, so then I went to a local one in Whanganui when I was still at high school. Got nowhere, but thought, this is a bit of me. When I went shearing with Keith, he'd done a few so he showed me, and his style was really good for show shearing cos it's all about the bottom tooth and making a good job but doing it as efficiently as possible. So I went to a few shows – it's a bug – I never won one for years, but I made the odd final, it just kept you going. By the time I got to the Open, it was a tough era, there were lots of good shearers – there was David Fagan pretty much winning every show, then there was Paul Grainger or Paul Avery and Alan MacDonald and all those guys from the King Country, so if you were making the semi-finals for a couple of years and getting in a couple of finals every year it became a bug: you're really trying to drive yourself to get better. It was a challenge: you work all day practising, trying to shear as fast as you can and as clean as you can at work, so that on Saturday or Sunday you could go and be competitive – have a go, see how you went. It was good.'

Bart and Nuku met at the Taranaki Shears in Stratford. Nuku's brother Ronald was competing alongside Bart and happened to make the introductions. The following weekend was the New Zealand Shearing Championship at Te Kūiti and Bart asked if she wanted to go up. Nuku laughs with the memory: 'He had a ride up with Reg [Benjamin] and he said, "Do you want to come for a ride?" And I said, "Okay then," so we went up to the New Zealand Shearing Champs and had a great time, and then they all got drunk and I ended up having to drive all the way from Te Kūiti to Stratford.' But the unplanned stint as driver obviously didn't deter their blossoming relationship.

With the commitment of a farm in Taihape and a large mortgage, there wasn't time to develop competitively. So Bart competed in the shows close to Taihape and would 'maybe go to the Golden Shears or New Zealand Shearing Champs, and if I was down south shearing – when I was young I was in the Intermediates and Seniors – I'd go to the ones down there . . .

'I love competing . . . it's doing the best you can. You can't worry about what your mate's doing, you've got to make sure you've got all your gear prepared. Then you're prepared and you get up and do the best you can with the sheep that are in front of you, and if you're good enough you'll make it through to the next round, and the next round, and that's the challenge. If you're doing real good you'll come out on top of the heap at the end of the day.'

While Bart says he only ever won one Open, he's happy simply to be involved:

'I don't mind being knackered at the end of coming fourth in the Open when the other eight fellas that made the semis or didn't even make it to the semis are pretty gun shearers themselves and you're right up with them; that's quite a neat feeling in itself.'

In 2001 Bart and Nuku made the huge decision to pool resources with Nuku's siblings and their partners to lease and eventually buy Mangaroa Station in the Ruakituri Valley and Ruakaka Station in Tiniroto. Within seven years the families had worked hard enough that each family could go out and farm on their own account. Bart and Nuku took on Mangaroa Station. Throughout those years the whānau shore all their own sheep, which enabled them to save one of the big expenses. It's the shearing industry that has helped them get to where they are today. Bart doesn't wield a handpiece as frequently these days, but he jumps at any opportunity to share his knowledge with enthusiastic novices who come and work in the station shed. One of his many protégés is Catherine Mullooly, who is doing well in Senior shearing competitions currently. He is also coaching two of his own children to compete: Atawhai and Ariana.

Bart's enthusiasm and drive are also channelled into giving back to the industry. He and Ronald, Nuku's brother, took on the role of running the shearing competition at the Wairoa A&P Show after moving to Mangaroa. Bart and Ronald had gone to compete their first year in the area and discovered the competition wasn't running as there weren't enough entrants. Nuku explains: 'Bart and Ronald said, "We'll take up the mantle."' So for the next twelve months we got sponsorship, and Bart rang round a lot of mates, said, "Hey, we've got this show up and running up in Wairoa again."' Bart and Nuku have been involved with running it ever since.

As tends to happen, one thing led to another. Bill Gaskill, the competition judge, then roped Bart, Ronald and Eugene, Nuku's other brother, into doing their judge's badges. Bart and Ronald completed all the training and Ronald judged at the world championships in Invercargill in 2017, and Bart has now taken on Bill's job as examiner for the East Coast. He is in charge of organising the judges for the East Coast shows and facilitating training days for the judges. He freely admits it's different being on the other side of competing. 'Judges give up a fair bit to make it all happen, too, and you don't realise that until you actually become part of it; as a competitor you've no idea what's going on behind the scenes.' Bart has judged at both the Golden Shears and the New Zealand Shearing Championships, and has been selected to be the New Zealand judge for the world championships in France in 2019.

Recently he was successfully nominated to be a world record shearing judge. He has undergone two years of training and will now hold the position for twelve years. Bart sees the appointment as an honour and is enjoying the role. He has now been involved on all sides of the records process, having been in a record-making team, supporting mates behind the scenes in record attempts and now judging. Nuku comments that there are more tallies being attempted these days, so it's getting busier, 'which is a good thing for showing shearing in a positive light; it's really cool'.

When the opportunity arises, Bart still likes to compete himself, but now this tends to be in the speed shears: 'One whack and you're in and out,' he laughs, recognising there is not the mental and physical drain when you're shearing one sheep as compared to twenty. When they moved to Mangaroa, Bart recalls, 'I just did a couple on the East Coast and that's when I got into judging. I'd go and judge in the lower grades then and have a go at shearing myself in the Open . . . I was still competitive as: I was making the semi-finals and sneak into a final and then the wheels would fall off,' he adds with a wry laugh.

Bart reflects on how competitions affect the industry:

PERSONAL BESTS

DEC 2011 CATHRYN MULLOOLY - 205
DEC 2011 CATHRYN MULLOOLY - 304
DEC 2012 JACK ROBINSON - 506
DEC 2012 PETE CHILCOTT - 602
DEC 2013 SCOTT MAULTSAID - 205
DEC 2013 CAMPBELL ROADLEY - 217
DEC 2013 JACK DEVER - 360
DEC 2013 ROWLAND SMITH - 705
DEC 2013 MATTHEW SMITH - 713
DEC 2014 KANE BATY - 304
DEC 2014 ADAM MORTON - 501
DEC 2014 JACK DEVER - 506
DEC 2014 STEVEN PLAYLE - 509
DEC 2016 TOM HANSEN - 205
DEC 2016 ATAWHAI HADFIELD - 103
JAN 2017 ATAWHAI HADFIELD - 200
DEC 2017 ARIANA HADFIELD - 102

ABOVE Bart and Nuku Hadfield's woolshed is an environment where young shearers are encouraged to challenge themselves. Even world record shearer Rowland Smith features on the station tally board.

RIGHT A family affair: Atawhai, Bart and Ariana shear their station lambs while Nuku works as the woolhandler.

'When I started shearing in '86 or '87 it was probably in its heyday; there were shows just about every weekend or two, or there would be two or three some weekends from October through till late April. Over my time the sheep numbers have obviously gone down, the shearers are overseas – a lot more going overseas – the people to run the competitions are not there and the sheep are not there, and it's dwindled away. There's a hell of a lot less than what there was. But probably the profile of shearing has gone up because of the media, and with the way these fellas are portraying themselves in positive ways it's done shearing the world of good. Those young fellas, even if they're not going to shows, they read about Johnny Kirkpatrick and all that, and if they got the chance they'd go and work with him; and that's bloody neat – that's what we want to have happen. Then hopefully they get the bug for those things and that helps the shearing industry out even more again.'

He is keen to see more young shearers competing:

'To me every Junior, Intermediate, Senior . . . if the contractor is on to it he'd give them a day off, tell all the cockies, "We can't shear on Saturday, all those young fellas are going to the show." Because it's doing them the world of good cos they're going to shear as fast as they're normally going to shear and they're going to shear it properly cos they're going to learn about second cuts and a good finish.

'When you come to the Open and you start smacking out four hundred a day, if you want to win shows you've got to show-shear them all. That's why Johnny Kirkpatrick and Rowland Smith are representing New Zealand: they shear every sheep like it's in a show, and that's what I aspire to do . . . it's physically and mentally demanding cos you are trying to shear it perfectly, every sheep . . .'

Bart is a brilliant advocate, not only for competitive shearing but also for the industry in general. He is keen to let anyone know what can be gained by taking on shearing:

'I reckon there's lots of people out there that haven't been exposed to shearing that would love to be shearers if they knew the lifestyle and what we got up to and how we travelled around and saw the world . . . I've got a lot of mates that make eighty to a hundred grand a year and they only work nine months of the year and travel all around the world – pretty good lifestyle and own a house and do all that, and there's plenty of fellas that go to uni that've got a lot less than that.'

Listening to Bart and Nuku talk about their experiences in the shearing industry, you hear their complete belief in its value and worth. As Nuku talks with warmth on what the industry has given them, Bart nods in agreement. 'It's something that you can do – you can do it as a career, a sport, but also as a family. That's probably a big thing for us: that we did spend so much time as a family and were with extended family. Our kids have been brought up watching their parents and their aunties and uncles, and it's instilled an awesome work ethic in them; it's become a natural thing because they've been brought up with it. I'm really proud – it always feels good when we are all in the woolshed and you see everyone working hard and sweating and you're thinking, "Yeah, this is a nice place to be" – it's honest.'

While Bart may have pulled back from competing early due to commitments with farm ownership, blade shearer Tony Dobbs retired from competitive shearing in the 1990s to take up farming – only to return to top-level blade-shearing competition in 2014. He has since represented New Zealand in blade shearing in two world championships.

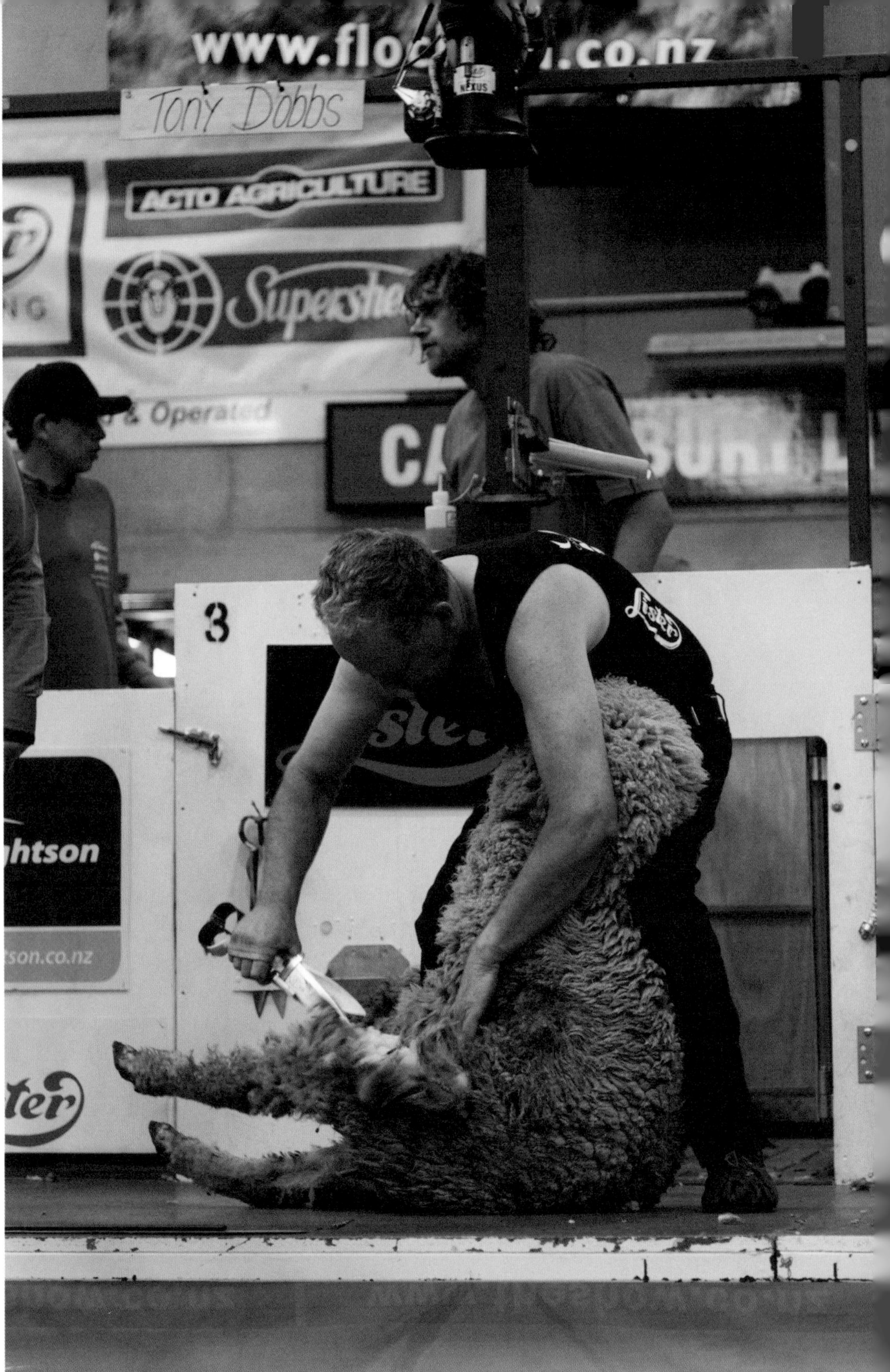
Tony Dobbs
ACTO AGRICULTURE
3
Lister

Tony Dobbs – Partying with the prime minister

*Peter went through the door first, and it was Prime
Minister Jim Bolger there, and Peter said, "How you
going, Jim?" And he says, "How you going, Peter?"
And I thought, it looks like we're all on first-name basis
so I'll call him Jim, too: "How you going, Jim?" "Good to
see you, Tony; help yourselves to drinks over there and
something to eat" – and so we did that all night. Met all
the hierarchy . . . oh, there was a whole heap of politicians
there and it was a good night. About four o'clock in
the morning Jim Bolger said to us, "I think you guys
should go home." I think we overstayed our welcome.*

As Tony recalls his trip to Waitangi sometime in the distant past
he laughs. When he and offsider Peter Race agreed to giving a
blade-shearing demonstration at the Highland Games at the Waitangi
celebrations there was no thought of partying with the prime minister.
There are many stories as Tony looks back over his career: like the time
in Kruger National Park in South Africa, when Mana Te Whata swam
in Crocodile Creek, so named because of its inhabitants; or the time Peter

and Tony convinced a TV producer that Prince Charles was going to give them a lift to Wellington on the Royal Andover, and the producer parked alongside the plane at Kerikeri Airport in the Holden ute only to get short shrift; or, in Tony's younger days, there may have been an incident with some pigs and fluorescent orange paint; and an old dray outside the hotel in Ōmarama may or may not have been towed to a new location under the cover of darkness. Oh, and then there was the rite of passage for newbies at Hakatere Station . . .

'It was in the mid-eighties . . . we used to have a thing called "the race of the flaming arseholes" up there. And this is for all the new boys that would join the shearing gang for the start of the season. Cos usually drink was involved and the Hakatere quarters were quite big quarters, they'd be like a big school. We'd wait till about eight o'clock at night, when it was quite dark, and the routine was they had to jam as much toilet paper as they could up between the cheeks of their butt. Then it would be lit, and see who could do the most laps of the shearers' quarters. Of course, one of us would be at every corner with a bucket of water if they needed to be doused. It was hilarious, and the atmosphere of the flames going streaking around the shearers' quarters at night was . . . interesting.'

Tony, to list just a few achievements, has been a world champion blade shearer (1988), is a Master Blade Shearer, and holds a Guinness World Record for the fastest blade shearing of a sheep. After coming out of blade-shearing retirement in 2014, he has faced-off with the world's best at the world championships in Ireland in 2014 and in Invercargill in 2017, placing third and second respectively. Shearing and competing have opened doors, and Tony is grateful for the opportunities the sport has given him.

When asked how he became a shearer, Tony quips, 'Probably by accident, to be honest.' He finished school at the beginning of sixth form, when he grabbed a building apprenticeship that came on offer. Unfortunately, soon after, there was a downturn in the building industry and he lost his position. So he went back to the family sheep and beef farm just out of Fairlie in South Canterbury. While working there, he thinks around September 1979, one of the local contractors, shearing legend Paddy O'Neill, approached him about rousieing for one of his gangs.

'I was picked up in an 850 Mini – a wee red Mini – and I didn't even have a sleeping bag in those days, so I had to take some blankets and sheets and a duvet with me and a little pack and away we went at about a quarter past six in the morning. I wouldn't have a bloody clue in the dark where we were, didn't even know where Tasman Downs was, and when the sun was finally shining in the light of day I could see we were beside Lake Pūkaki looking over at Mount Cook. And it was quite a small, wee pokey shed. There were five shearers and I remember a little tiny window at the end looking towards Mount Cook – it would have been all covered in cobwebs – and, yes, seeing blade shearing . . .'

While Tony was familiar with the inside of a woolshed he had never actually seen blade shearing before, so there was a lot for him to take in. He remembers how he felt the first time he held a pair of shears:

'The first time I touched a pair of shears, to actually have a snip with them would have been in that shed, Tasman Downs, and I thought these things were amazingly sharp and would cut the sheep in half.'

To progress to owning his first set of shears, albeit a hand-me-down from a shearer, was 'absolutely magic'.

'It was like Christmas . . . I made sure they were wrapped up every night and put away so they wouldn't get damaged, because a pair of shears is a pair of precision cutting machines and you only have to bruise their edge and they won't cut wool. So they'd be wrapped up and put away in me bag and carted away home at the end of the week. If I wasn't using them for a while I'd put some oil on them to keep them right. And we also had to put a thing on them called a cockspur, which is a wee piece of stainless steel at the back of the shears, so that when you open them too wide they don't fold over the wrong side and cut your thumb. I think I went and got an old sheepskin out of the back of the shed and cut it out into a wee sheath and sewed it all up, so they had protection for going in the bag. I was probably told to do all those kind of things – "go and get that old sheepskin and get that . . . make sure you look after those", they would have been telling me, and I of course would have said "yip, yip". And I did the same thing many years later for a young lad that was keen. Gave him a pair of shears to look after. "You looking after those shears?" "Yeah, yeah."'

A blade-shearing apprentice goes through much the same steps as a machine-shearing apprentice: rousieing and pressing, and then on to finishing the last side and onwards. The tallies are slower in coming, though, and they don't go as high as in machine shearing. Within three years a blade shearer aims to shear around 140–150 sheep, moving up in increments to 150–160, and finally aiming for 180–200 as they gain experience. Tony thinks there are only a dozen or so blade shearers in the country that have ever cracked the 300 mark; he is one of them, having reached his 200 and 300 tallies in the same day, along with John 'Boots' Waters.

Reaching your first 100 is a significant milestone for a blade-shearing novice, and Tony still remembers how it happened:

'We were shearing at Richmond Station, which is on the side of Lake Tekapo. They were short of one shearer and they were on merino wethers – quite big wethers, too – and the contractor said, "Instead of pressing, do you want to be the shearer and I'll get someone else on the press?" So I shore, I think it was for three days. I think I was doing about sixty, sixty-five a day on these merino wethers cos the back was hurting and a sore hand; yeah, they were big boys. But I never gave in: I got to the end of the week and got the job done. Then about a month later, we were at Blue Mountain and they were short of a shearer again and they gave me the opportunity to have a go. I started on the merino wethers, and on the fourth day we started on the hoggets and I shore my first hundred. That was before I'd even been to a shearing school or anything, so I was quite happy, and the tradition is when you shear your first hundred on the blades you've got to shear the hundredth one in the nude. Tradition had to stand that day, and I did that. Then you've got to shout for everybody. In those days it was a couple of crates of flagons, if I remember rightly, on the Friday night – so you probably shore all week for nothing by the time you paid for the beer. But it's something you'll never forget. I still remember over the next twelve or thirteen seasons when I was shearing, all the boys that shore their first hundred doing the same tradition, and some sheds have got photos of them along their walls all doing their hundredth sheep.'

Tony doesn't recall anyone slipping with their shears on those occasions.

Keen to save for a farm, Tony took up machine shearing as well so he could work a full year, but he always preferred the blades.

'Machines were more physically demanding. Yeah, I found it a lot harder. When I was blade shearing I was around a lot of good shearers and I learned a lot of skills from them; I was around a lot of good machine shearers, too, but they were never camp-aways, so you don't do the old information pool like you do in the camp-aways when everybody's there together and talking about things and swapping ideas. I remember all us young guys, we'd be down there trying this, see if this works and fiddling around; after tea we'd shear a couple or do something with gear . . . trying to come up with something out of the box. I used to do some crazy stuff. But that was all about learning something that might work down the track that was a bit different again.'

Tony acknowledges the input of the older guys he shore with:

'The like of Tony Hands, John Kennedy . . . there's an old guy who was really good on gear and I took a lot from him, that was Maurice Oakley. Paddy O'Neill, Paul Rose – they all had their unique wee thing that was – maybe they didn't have the whole package, but they had some particular part of the sheep where you thought, "That looks good, I'm going to work on that and fit it in with the rest of the jigsaw puzzle."'

Honing his skills further came through competitive shearing.

'I think I got talked into it by Paddy O'Neill . . . he was quite keen on the young lads displaying their skills or abilities at a competition, and probably it was good for them socially; or, presenting themselves in that way, they could talk better to farmers' wives and the like – a bit more pride in themselves, I suppose. Went to me first one, would have been Waimate Shears . . . and I shore in the Intermediate, but I was shearing as a first-year learner. It was the first year I'd been shearing and I was wanting to get the first-year learners' prize, which might have been fifty or sixty dollars. Anyway . . . I ended up getting into the final of the Intermediate, and cos of not knowing how the things worked I was still trying to work out whether I'd won the first-year learner – which I had. I ended up getting third in the competition. It was the buzz you got out of it, the mental buzz. I was very nervous the first time, I was absolutely shaking; I remember my leg kept jumping up and down off the floor when I was going up the neck of the sheep – terrified the sheep would get up and run away or do something silly. But when it was all over and done it was meeting all the other competitors that you shore with – you're meeting them and getting to know them from different gangs and learning how to be competitive. And I got hooked very early.'

And he well remembers his first big show:

'It was the Christchurch show, which is the New Zealand Golden Blades, and I was shearing in the Open. I think it was my second year, so I had just turned twenty. In the first year I got into the final – I think I got third or fourth – and I thought, if I made a final in the Open that was good just to make the final. Then I got into this final and I shore very, very cleanly. I wasn't fast – I was probably a couple of minutes behind the first guy off – but me job pulled me through and I won it at twenty. And for such a young guy to win . . . it made all the papers and God knows what . . . And I thought, from then on just making a final wasn't going to be good enough for me! I changed my attitude a wee bit; I was out there to win.'

As he started out, the young man was a little awestruck by some of the legends of shearing attending the shows.

'The big names in those days would have been John Fagan from the North Island, Colin King, Peter Lyon, Kevin Walsh, Roger Cox, Adrian Cox . . . they all had the big flash cars, and being a young boy you'd think, "That's me – look what they've got." It was the image of it all then, the Te Whatas, Samson and all them. You'd only make eye contact cos you were only a boy, but after a couple of years they'd seen you at the competitions and they'd be saying hello to you and you'd be thinking, "Jeez, he knows me!" That was a bit of a buzz. But at the end of the day, they're only human; they're no better than anybody else. I got to realise that pretty early in the piece, and you're only human too, and don't think you're better than them or one class above them.'

Preparation in the early days was simple:

'It was just a good pair of shears that were going that week; you'd put them aside on Friday and that was your competition pair. That's what I did, and some of the guys still do that today. In the later years . . . early in the year I'd find a pair of shears – for some reason they stood out to me, I thought they were something a wee bit special – and instead of shearing with them during the week I'd put 'em aside very quick and start on another pair of shears and they would become a competition pair. And I'd always have two competition pairs – so, same thing, as the season went through I'd find another pair that I thought stood out. And I'd make sure both pairs were identical in every feeling when you shore with them. I mean, if I shut my eyes, no matter what pair I pulled out they felt the same to shear with, so you're totally confident with them.'

With a grin and a laugh Tony reveals the true secret to his success:

'The first time I won the Christchurch show I had this pink towel and I thought it must have been my lucky towel. So, every competition I'd have this pink towel, just a wee towel, and take it up. I remember one time I accidentally left it behind and I didn't win the competition, and I thought, I never had my pink towel – the word got out . . . It was just psychology. It was a little mental game thing that I played with the other guys – "I've got this lucky towel." Of course, everybody threatened to burn it.'

All joking aside, Tony explains more of his approach:

'As you got more experienced, you got more prepared. You took all those little bits of the equation that could go either way and you made sure it went your way with your gear. You'd never go half-prepared, because you couldn't afford to; when it got down to winning or losing by a fraction of a point you had to have everything right – even the way you opened the catching pen door, or the way you stacked the sheep into the pen. As to studying other shearers, I studied David Fagan and a few other guys. I thought, that guy could get the sheep out and the belly off before anybody else got out of this pen – what's he doing? Went up and studied him and worked out how he was doing it – never said nothing. I knew how to do it then – and I did it for many years, and I'd be the same: I'd have so many blows into the sheep before anybody even got their sheep out onto the board. It's only a two- or three-second thing, but I've lost by half a second or won by half a second over my career, so you stack all those things in your favour and it makes a big difference.'

Tony rattles off the opposition in his early days: Peter Race, John Kennedy, Paddy O'Neill, Paul Rose, Craig 'The Colt' Kennedy, Alex Sole, Peter Casserly, Donny Hammond, Billy Michelle, Ross Kelman. He explains how he approached shearing with them:

'I was a great one on reading my opposition before I shore with them: knowing what I thought was their weak point and trying to get that to play against them by what I did . . . well, if it didn't play against them, in *my mind* it was playing against them, and it made me shear more confidently, was maybe more the word. The hardest ones I found to shear against were the ones I'd never shorn against before, because I didn't know their weak points. I'd study them early in the day and see if I could see any weaknesses in their system, or whether their nerves were a problem, and work on those things – maybe come out and shear one real fast and get him excited, or are you better off just going slowly? Or doing your roughest ones or your hardest shearing sheep first, and slowly getting faster and faster. It's just working out which way to attack the opposition. And hopefully they're thinking about you, cos if they're thinking about you and not about what they're doing they're making mistakes. That was my idea, and I was only thinking about what I'm doing: I'm doing this, and this is going to do this to them – away I go. And if it worked, that made me feel better, and if it wasn't working, think of Plan B. There's always a Plan B.'

And Tony smiles, perhaps with thoughts of when Plan B was thrown into action:

'I remember one particular competition – it was the Open final – first sheep I went up the neck, and something went wrong and I poked the sheep in the neck. I knew I'd done it and I thought, "God, if this is rally driving I've just rolled the car, what am I going to do?" Luckily it happened on sheep one, so I knew I was going to get a maximum score on that sheep cos it would have to have a stitch put in its neck. So I just opened the throttle right up and went flat out – I think I was shearing them in under two minutes, put about a sheep and a half round everybody – and waited till everybody finished and walked off. And I thought, "Well, that's my night gone." But cos I shore the other five immaculately at fast speed, I still won the competition. I couldn't get over it. I thought, "There you go, I didn't chuck the towel in; I worked out what I had to do and changed the disadvantage into an advantage for me." If it had been sheep five it would have been too late, but sometimes these things happen. Make the best of a bad situation.'

One technique Tony has honed is to mentally shear the sheep before he has even started and to mentally eliminate his opposition.

'I'd looked over the pen door and I've mentally shorn all those sheep that are in that pen before I shear them. I know how they're going to shear in me mind, and I know which one by the look of it might give me a bit of trouble. For some reason I know which ones are going to be the troublemakers just by looking at them. So you're prepared for anything that's going to go wrong, because you know things happen during a competition. I used to say to myself, "If I do this and this, this will put pressure on the guy on the left-hand side of me." And by sheep three if I'm in

The importance of sharp shears, particularly when competing for a position in the New Zealand Blade Shearing team, cannot be underestimated. Tony Dobbs, Waimate Shears 2017.

front of him I put a wee cross through his name, forget about him . . . and so I was mentally wiping them off the board before I shore them and away we'd go. And that guy down the end, he's going flat out, and he shouldn't be going that fast cos I know he'll be getting rough, so there's a wee X that goes through his name . . . but I'm still worried about the young guy on stand one because he's not far behind us and I know he makes a good job; we'll put a bit more gas on and hopefully that makes him rough. So you're playing mind games, making yourself comfortable with what you're doing.'

The 1988 Golden Shears in Masterton where Tony competed in the world championship was his first real experience of feeling the pressure and he realised he had to get his head around it. In South Africa the following year the pressure was even more intense:

'It was the bloody national anthem, actually, that's what got to me: just standing, listening to your national anthem with the New Zealand singlet on, and no, you're not shearing for yourself. When you're in a competition you're shearing for your own personal pride or you're doing it for yourself, but when you put the New Zealand singlet on you're actually shearing for four and a half million people. It changes the equation a bit, pumps you up an extra level – and of course the national anthem makes it even bloody harder. I wish they wouldn't do that, to be quite honest . . . I was just about teary-eyed. Later I learned to switch off before all that happens, and you go into another little wee world and probably become a zombie. You could look out at the crowd and you don't actually see one single person, nor hear anybody call your name – you just turn right off. I learned how to do that, and I've tried to teach other people how to do it.

'At the world champs this year just gone, in Invercargill, one of the Scottish lads I'd been working with the previous week, he was already up in what they call the holding pen – it's where you're all lined up for the next two events away. I saw him sitting there and I could see the tension on his face, and I said, "Okay, who's got the lollies?" I had the bag of lollies – "It's time for a lolly. Okay, jokes time." And we just start talking and telling jokes and eating lollies – and I said, "The green ones have got the most go; get into them." Just trying to take his mind off it – "Okay, righty-o, you got just over a minute to go," I said, "now you can have some think time and go out and do it." But I said, "Don't start doing it ten minutes before – you'll get too hyped up with ten minutes to go." I said, "Look at all that nervous energy you were burning; just have a bit of fun time." And I did the same with him the next day: I said, "It's your turn for jokes cos I've sort of used all mine." And I said, "Just turn off when you get up there, there's no one looking at you. Tell that TV camera to buzz off, get out of my face, if he comes in." I never really get nervous any more; I don't let the moment take me away – just think about something else.'

Ultimately, Tony's focus had always been on buying his own farm: 'You had to have an ambition, otherwise you'd just spend it on fast cars, drink and other things.' Years of daydreaming about owning a farm and combing for the right one as he covered the miles from woolshed to woolshed eventuated in Tony and wife Julie buying their Cricklewood property in 1990. He shore part-time for a while after the purchase to supplement their income through difficult economic times, and carried on competing for a number of years. When he retired from competitive shearing he had a pile of ribbons and trophies to his name.

Tony started judging shearing and was an instructor for Tectra, so he stayed involved in the industry, but never thought he'd be competing again. 'I was sort of bullied into it – I didn't really want to come back and I've been poking around ever since.' Tony goes on to reveal that John Hough was the bully. He was competing in every show around the country before he turned seventy – or 'Houghy's Last Stand', as it was known – and John worked on Tony, saying he should be competing in the world champs in Ireland. Tony was diffident but agreed to compete at Waimate and, as he says, 'it went from there'.

It was the South Africans, Mayenzeke Shweni and Zweiliwile Hans, who beat Tony in Ireland. Not until after he beat them in the individual shearing and in the team event with his partner, Brian Thomson at a tri-nations in Australia was Tony confident that he could get on top of them. It was the encouragement he needed to begin his campaign to fight for a place in the 2017 world championships, which New Zealand was hosting; having won every competition since 2014, he felt he had earned his place in the team.

'I went down and thought, "Right, I'm ready for it: bring it on and we'll see how we go." The sheep were very, very feisty, but first round they sat well for me and I loved them, they were beautiful to shear, and I top qualified again. I was about seven points in front of the South Africans and I thought I'd got them on the back foot a wee bit. But I thought, "Think about Ireland" – same thing, I was miles ahead and they seem to know how to come back when it's crucial. And lo and behold, in that final there he did come back, and I thought I had him, too.

'On sheep three I was in front and I knew I was shearing them well. We were shearing full wool for a kick-off, so we had to shear three full wool and then three second shear. And these second shear were crazy. They were cross-bred – they were fired up like they were ready to go to the ram, they were real feisty. And I thought, "Right, when I go in to get the first second shear, bring it out quietly and sit her down, don't rattle it." And I brought it out and sat it on its bum gently and then started shearing it. I slowed down, and she's sitting, and I started winding into it, she's sitting, she's away, right down the hole. Went in quickly for the second one, pulled it out and it was jumping. I felt like letting it go and putting it back in the pen and getting another one. It would not sit: it jumped around, stepped up the neck, do the neck, it threw out at me, behind me, stepped up again, it threw forward. And I thought, "This thing ain't going to sit." And then I thought, try and get it on the other cheek of its bum. So I finally got the neck done, fought it all the way through the first shoulder; by that time he'd got out in front of me . . . And I thought to myself, "Don't do anything stupid." So I shore it, I would have lost a job on that one a wee bit, and then I got the last one out and I thought, "Just bring it out quietly and don't upset it." And it sat, and I started to catch a wee bit of lost ground then, and buttoned off and looked at me card that the judge had – happy with that – but in my own mind I thought I'd get somewhere around third. I was thinking the two South Africans might have got me.

'When the final result come through, I'd done the best job out the back and on the board, and it was just time that caught it. I kept thinking, "If that sheep had sat on its butt . . ." – but it's all part of the game. You just got to let it go. But I still see that sheep, feel that sheep, and think, "You little madam." He might have had kicking ones, too; I wasn't watching what he was doing. I was just doing what I was doing, and I knew what I had to do and that was to put them out quick and clean. I put five out of six quick and clean, but five itself didn't want to sit on its butt . . .'

Tony talked to Alan 'Mickey' MacDonald, a world champion shearer in his day, who was working out the back at the competition; Alan was sure that 'these sheep are going to make grown men cry'. This draws a wry recollection from Tony:

'I thought about it that night after the result, and I thought, well that particular one just about brings a tear to my eye, the sod. I'd like to see it with mint sauce on it.

But anyway, that's all part of the game and you can't let it get to you . . . when you get a kicker you just have to suck it up and somehow get it off without it throwing you off your stride too much.'

To compete in New Zealand meant a lot to Tony and it 'hurt' not to win.

'A lot of people thought I could have won it . . . we won it last in '88, then come back for Ireland in '14, and win it in 2017 – that would have been the fairy-tale ending. And some of them picked me to win it. The way I was shearing earlier in the week I was going to win it – I wasn't paying much at the bookmakers, anyway.'

Despite the pain of being pipped again by Mayenzeke, Tony is ever philosophical:

'I said to meself when I come off, "Well, I didn't do anything stupid; I haven't lost because I've done something stupid like cut a sheep or do something that I don't normally do." That's when it really hurts you, if you do something silly – if you snip a bit of skin off . . . I was happy with what I'd done, the way I did it; it was just a pity that feisty wee thing didn't want to sit quiet.'

Not one to give up easily, Tony is in the throes of competing for a spot at the world champs in Le Dorat, France in July 2019. He appreciates all the opportunities that shearing has given him: three trips to South Africa, competitions in Australia, a world championship in Ireland, travel around the United Kingdom, and opportunities to meet people from all walks of life and make lasting friendships. The farm that Tony and Julie own was bought with shearing proceeds as well. None of it, of course, just dropped in Tony's lap: it all comes from doing the hard yards. It's fair to say, though, that there would have been little thought of such possibilities as the young seventeen-year-old lad headed off in a little 850 Mini to Tasman Downs Station.

Tony Dobbs is not alone in coming back from retirement and challenging for a world title. The legendary Brian 'Snow' Quinn, who retired from competitive shearing in the early 1970s, came back and won the world championship at Masterton in 1980. It was one of many high points in a long and impressive shearing career.

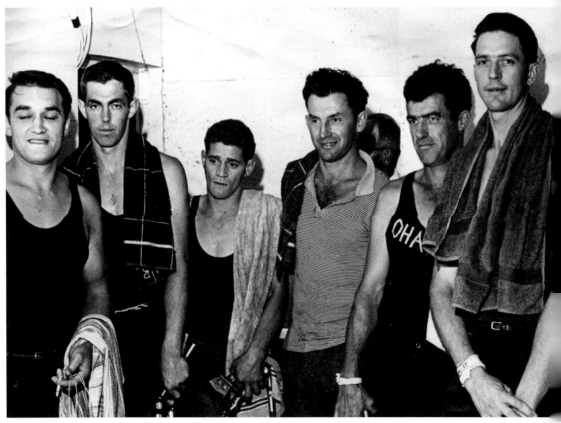

Brian 'Snow' Quinn – A competitive shearing legend

'Snow Quinn, he won six Golden Shears; he's won the Alex show three times, he's recognised as New Zealand's gentleman shearer. Keith Holyoake said he should have been a prime minister.'
– Peter Casserly

Standing a little over six foot three and weighing in at just nine and a half stone, Brian 'Snow' Quinn lacked the ideal build for a shearer. He was considered too light, and people thought he wouldn't have the requisite strength. Brian, of course, was to prove the naysayers wrong, becoming one of New Zealand's legendary shearers: six-time Golden Shears winner, a Caltex National winner, world champion, four-time NZ Merino Shears champion, nine times Southern Shears Champion, in winning teams in the worlds and the trans-Tasman competitions, and a Master Shearer, as well as being awarded an MBE for services to shearing. Also, the unassuming Snow Quinn must be the only shearer who can attest to having both a song written about him and a racehorse named after him.

When you listen to Brian you get a clear sense that he's a keen observer of people and a thinking man. Never having attended a shearing school – they were only just getting under way when he was starting out, and he was already busy working – he learned his trade by carefully scrutinising the technique of others and taking what he'd learned back to the shed to practise. 'Nobody did teach you much in those early days, nobody told you much; it was just look and watch.' One person to leave an early impression on the young Quinn was Malcolm Barclay:

'It's surprising what you learn at a young age . . . In those early days I used to do a little bit of work with Malcolm Barclay . . . he was probably the first good shearer who could shear three hundred sheep a day, every day, and even now I still think back to some of things I learned just from watching him. He was one of the first ones to have concave gear, for a start. Course, when I learned, I learned with concave gear. One thing that I've always remembered, if he didn't shear for a week he'd always regrind his combs and cutters because he said they never cut the same. And that's going right back to the late fifties. Oh, it was mainly the quality of his shearing: he was a very, very clean shearer and I think that always stood out . . . how clean his sheep were when he shore them and the numbers he was doing at the same time.'

Bing Macdonald, Joe Ferguson and Colin Bosher were fighting for supremacy at the time when Brian was developing, and he watched them with great respect. It was Bing's style that the young man tried to emulate, although he came to the conclusion that he couldn't copy him exactly as their builds were so different. Bing's slow hand and how 'he started his blows properly and finished them properly' were traits that Brian did take on board. He reflects on how Bing was 'a very, very good clean shearer'. And in those early years a consistent theme Brian heard was that you needed to start a blow properly and finish it; 'if you don't start it properly it won't go into the wool, your comb won't go into wool; and quite often if you don't finish it properly, it takes another blow to finish it.'

Opportunities to watch the guns at work came through attending competitions. 'I just got in a nice position where I could see them when I was starting off. I could watch them and see how they done it and then practise in the shed.' Brian is adamant that competitions are the place to be to improve your shearing. He admits to not knowing what a second cut was until he attended competitions.

'But then, going back to competitions, that's where you learn all these things – like, you're getting marked on your outside job, you're getting marked on your time, you're getting marked on the quality of your wool. So, if you go back and have a look at your points and your outside job's good and the quality of your wool's no good and your time's no good – well, you think, "Right, I've gotta get my time sorted and I've gotta get my quality of wool sorted" – that's the way I approached it. I shore in competitions from 1959 up until 2006 and I always used to look at the points to see how I was getting on and where I could improve. Every competition, you'd look at your points and see where you're losing your points, and it's pretty straightforward.'

Brian's illustrious competitive career started, like much about him, unassumingly. Growing up in Aria in the Waikato, he entered clean-shear events at local A&P shows in the King Country at Piopio and Aria in the late 1950s and early 1960s. Then in 1962, curious to see what this new competition, the Golden Shears, was about, he headed down to Masterton with some mates.

'I'd only ever shorn in one or two competitions. I'd never shorn in a Senior competition because they didn't have them in most of the shows, and the Golden Shears had the Senior and that's the only one I've shorn in in my life. I actually just went down to have a look, with no high expectations. But it was one of the highlights of my life, really, when I won it.

   'It's something about the stage . . . such a vast big place, and six stands – it's unheard of anywhere – and the atmosphere: three and a half thousand people there, it's like going to a rugby match. It's just overpowering . . . I was stage shocked, really; I've always remembered that, it's been a highlight of my life.'

After the surprise win in 1962, it is fair to say that the next two years were a time of consolidation for Brian. Competing in the Open class was a step up, and he also had to adjust to the North Island shearing conditions, having moved to the South Island in 1963. Used to shearing smaller full-wool sheep, he freely admits to being out of his depth when shearing the larger second-shear sheep of the North Island. Typically philosophical, though, his ousting in the semi-final in 1963 did not overly perturb him:

'It wasn't tough: it was just a fact of life, I wasn't up to it. It didn't worry me too much. That was when I realised what my limitations were, that I wasn't up to it; and that's why I had to shear more, and prepare myself better for bigger sheep.'

Preparation came through travelling prior to the show and getting his hand in with some North Island sheep.

'Most of the shows, you tried to get at least a day on the type of sheep you were going to shear. And the same at the Golden Shears. I went up to the Golden Shears ten days before and done a week's shearing beforehand on second-shear two-tooth sheep . . . I used to shear for a guy called Athol King at Pahīatua and some of the locals didn't like us very much because we were cutting them out of some of these two-tooth second shear. George [Potae] and I went and shore for Athol King for probably five or six years, and he'd give us a week, ten days' work. Which was very, very good. That was quite important to go and shear a few days before the competition cos we didn't have sheep anything like they had up there.'

And Brian returned to compete in 1965.

'I didn't think I would ever win it – I just enjoyed doing it. But winning it was a bonus. Oh, an unreal feeling . . . I'll never forget it as long as I live. I just couldn't believe that I'd won. I think it's the greatest feeling I've ever had in a competition, that one in '65.'

Winning the Golden Shears created quite the buzz in Alexandra; a crowd gathered at the aerodrome to welcome the champion home, and a mayoral reception was given in his honour. It was to be the first of six wins.

Remembering his 1966 final Brian reflects, 'I learned a hell of a lot that year . . . I was very disappointed with how I shore, but I was delighted I got third; I thought I would have been sixth for sure. I couldn't believe I got third after the way I finished the last three sheep. I was just absolutely buggered: I couldn't keep the comb on the skin.' With gear trouble, and using a borrowed comb, Brian struggled to get through his sheep. In hindsight he realised he should have gone back to his own gear, but recognises that it 'was just lack of experience not knowing what to do when things went wrong. And in 1967 I used the same two combs that I used in 1965 and there was no comparison. In '68 I used the same two combs, and in '69 George Potae used the same two combs.'

Brian recounts some more pearls of wisdom from Bing:

'I always remember what Bing Macdonald said to me: "Put your best shed comb on and go for it." For the final that's what he told me, and it always stuck in the back of my mind. You've got to remember those combs were never used anywhere else but for the final . . . I've still got them.'

Brian used a different set of combs for the heats, semi-finals and final. This was to do with the different styles used at each stage, as he goes on to explain:

'You had safer gear on for the heats so that you didn't make too many mistakes; if you made mistakes you never got in. And then as you progressed you had to shear a little bit faster and a little bit better, so you'd change your gear accordingly.'

Brian went on to win the Golden Shears in 1967, 1968, 1970, 1971 and 1972. While the accumulation of wins brought him superstar shearer status, it was the first two of these wins that Brian really rates:

'Winning in '67 was a good feeling because I felt a lot better and I shore a lot better and I didn't have any hiccups. Sixty-eight was the best final that I ever shore in the Golden Shears and that's the one I enjoyed the most. I never had a thing go wrong, I never had a sheep put a foot out of place, and I led in the final from start to finish. I never had one hiccup during the twenty sheep, which is very, very unusual . . . I felt quite comfortable. I still didn't know whether I'd won it or not, but I knew that was the best shear I'd had.'

The memories of those early Golden Shears have stayed with Brian down the years:

'It was amazing to be a part of those early ones. You always had a few beers and a few laughs, but you were meeting people that you only met at the Golden

Shears. That side of it was very, very good. Later on that changed, when the drink-driving laws came in . . . but in the early competitions it was always a very social atmosphere afterwards. They always had a dinner and entertainment after the competition and that was enjoyable. And you could hardly get in there for the amount of people all about; the whole town got behind it. Television was coming into its own; there was a lot of advertising on TV and in the paper, and I think it was mainly that the people in Masterton got behind it and pushed it that hard that everybody stood up and took notice. Anywhere else it might not have worked.'

With three Golden Shears wins under his belt, Brian had people suggesting that he consider no longer competing.

'Some of them were saying, "Why don't you stop shearing now and give the others a go?" That didn't make sense to me. I don't know why they would be thinking that, because anybody's capable of winning it if they're in it.'

As it was, he did not compete in 1969. Without wishing to lift Brian into the realms of sainthood, the 1969 Golden Shears stands out as showing the mark of the man. To support his son David in his competitive swimming development, Brian stepped back, freeing himself up to attend swimming carnivals, as well as supporting his good friend George Potae in going for the purple ribbon.

'I said to George, "You can win the Golden Shears this year." I said, "I'll help you." And he won the South Island Shearer of the Year that year, too. He used my gear all year in the shows in '69, but that was just for the way he was treating me. It was payback, really: he looked after me, gave me good sheds and put me out with good shearers and all that sort of thing. He wouldn't let any of these speedsters go out with me cos he said, "All they want to do is beat you. It's not worth the effort." And he didn't like me going with some of the younger top show shearers, cos he said, "They're only going out to learn off you to beat you." Which is probably true. But that's life, isn't it? That's the way it goes.'

Brian also remembers how George prepared for the competitions:

'He used to shear for me Friday, Saturdays and Sundays and I'd get paid for them. I'd come home . . . he'd take my stand and go and shear and put those sheep down for me. How many guys would have done that? Not many. He even came up here one day and shore for me.'

Brian would set up his gear for George for the shows, but everything else was up to George. He also may have had a part in coaching George in the art of brinksmanship, and he laughs as he recounts the lesson:

'To qualify for the South Island Shearer of the Year you had to be top qualifier at the Gore show . . . and Joe Ferguson was well in the running for it, too. I was down there – I didn't shear, but I was down there – and Joe was very disappointed with the way he shore. He qualified well up, but he didn't qualify top. And I said to George, "All you've got to do is get up there and think you're in a final and shear a minute quicker than Joe, same job, and you'll have him." Which he did: he shore like he was in a final and he top qualified. See, he was top shearer. The following day, Joe called in to see George's brother in the office, and Mac said to Joe, "How'd you go, Joe?" "Oh," he said, "those pair of bastards, they put it across me." Like, he knew as soon as he finished we were on the case – he didn't shear near as well as he'd like. But you've got to take advantage of those sorts of things.'

Brian stepped back into the limelight in 1970, winning the competition that year and the following two. With the needs of his family changing, and feeling that he was not as fit as he once was, he retired from competing nationally. Each of his Golden Shears wins had reaffirmed his place in the shearing hall of fame. He had truly left his mark, and his wins were to leave him a household name and an idol in the eyes of those young shearers coming up behind him.

Brian's humble nature comes to the fore when he is quick to acknowledge those who enabled him to become a Kiwi legend. Firstly, he openly admits that without the backing of his wife, Lyn, he could not have competed. And then there were the contractors: 'I've been very, very privileged over the years; the contractors I've had have always looked after me,' and George Potae and Murray McSkimming earn a special mention:

'You've got to realise that when you're going round the competitions and you're working for a contractor you can't just go where you like; you've got to go where you're told, and if they took a snitcher on you they could send you where the sheep weren't suitable all the time. They always used to make sure you got suitable sheep and always gave you time off and things like that. So I was very, very lucky in that respect with everybody I worked for. They always gave me decent sheep to shear before a competition . . . and if it hadn't been for people like them I wouldn't have had the opportunities to go in a lot of the shows.'

As if being six-time Golden Shears winner – a feat surpassed only by Sir David Fagan – were not enough, Brian took on the mantle of shearing world champion, despite not having competed at national level for eight years. The world championships, held in conjunction with the Golden Shears in 1980, was newly established in 1977. Brian is typically low-key in his response when asked what brought him out of retirement:

'I was still shearing then, I was shearing full-time. I thought I'd never get another opportunity to do it – that would have been my lot. I just thought I'd have a go at it and see. I was very, very lucky to win that cos I was past my best by then and I had too many other things to do. I couldn't really settle down and concentrate on what I wanted to do in the shearing side of things cos I was running a gang and organising things and all that . . . And my family were growing up then, too.'

Further recollections are more on how badly he felt he shore than on becoming a world champion:

'But I was very disappointed with the way I went – in the way I shore, particularly . . . I wasn't shearing well, I couldn't get at them. I didn't shear near as well as I'd liked to have shorn. If you shear well and get beaten – well, that's just the way it goes; but if you don't shear very well and get beaten, you could do better. That's my philosophy. I didn't feel as though I shore as good as I could shear, and I shore one sheep too many with the cutter I had on, which didn't help. I just wasn't comfortable all the way through. See, we had two qualifying rounds, too; I was way down in the first one, and I was still on the back foot after the second one – even though I did shear a lot better.'

There was pure relief for the man from Alexandra when the final was over, after tightly fought action between team-mate Martin Ngataki and himself. Observing Martin keenly in the team event, which Martin and Brian won, enabled Brian to pip him at the post in the individual final.

'Martin Ngataki was a good shearer, too, and he was a nice guy as well. We shore in the team, and I shore first and then he shore, and I said to him, "Put a bit of pace on, Martin," and as soon as he shore fast his second-cut rate doubled or more. So I thought, in the final the only way I am going to get him to make mistakes is to make him shear fast. If I let him do his own thing he'd beat me hands down, cos that's the way he was shearing. So I almost put a sheep round him, although not quite – generally if you can put a sheep round somebody it's pretty hard for them to come back. But I wasn't quite a sheep round him, not quite, and he almost won it. Very, very close – point three of a point.'

With all of Brian's national and international success, it is easy to lose sight of the fact that for forty-nine years he shore hundreds of thousands of sheep in many sheds around Central Otago and Southland. He had many more years quietly going about his daily work than he spent in the glare of the spotlight – although the spotlight's glow has never really left him.

Brian 'Snow' Quinn was born in New Plymouth in 1941, the fifth of seven children to William and Mary Quinn. His early life was spent on dairy farms, first in Urutī, Taranaki and then in Rangiātea and Ōtorohanga. His father was a sharemilker until he was able to buy a farm in Aria, when beef and sheep became the focus. For Brian schooling was something to be endured; the outdoors held far greater interest. A highlight of life at Ōtorohanga was the pursuit of wild turkeys – 'I used to love chasing turkeys when I was a kid' – and when they could run no longer, they were taken home to eat or possibly breed from. The 'free-range' life won over the classroom, and he laughs with the memory of his running prowess:

'I spent a lot of time running in those early days . . . running away from school, running away from the bus . . . I never liked going to school for some reason – I don't know why. I was very shy, I wouldn't talk to teachers . . . that probably didn't help . . . not so bad with people I knew, and as I got older I improved.'

By fifteen Brian was working – secondary school was not for him – and he still remembers his first job in a shearing shed. After probably only five weeks there, he was asked to take a stand when a shearer didn't turn up:

'They asked me if I'd like to shear and I said I'll try – have a go. We were shearing for Mervyn and Collie Watson . . . and we were shearing lambs, and I shore 139 that day. I wasn't very tidy, I don't think, either, because every time the farmer came in he'd say, "I'll show you how to do one," and I'd be watching him and I'd think, "Gosh, I wish he'd hurry up and get rid of that lamb so I can shear another one." I've never, ever forgotten that day.'

Brian's not too proud to admit that the first lamb he shore was none too flash. 'I don't think you'd like to put him in a competition. The most difficult parts on a lamb is around the head and the ears, and that's the first part you look at when you see a lamb standing up, round the head and the ears.' Considering he shore 139 on his first day when the guy next to him shore only 151, it seems that Brian acquitted himself well, and it was perhaps a sign of things to come. From that day on he knew what he wanted to do. 'I just loved it once I started; I don't know whether it was the challenge of getting the wool off, or the company in the shed, or the camaraderie of everything that's going on.'

Dribs and drabs of work followed, but the next big opportunity came via a family friend.

'The first real job that I had – like, shearing – was helping a guy, McKenzie; he was a friend of my father's. His boy was away doing army training and he had three thousand ewes, and I went and dagged all the ewes and crutched the lambs. Then the shearers never turned up, and he brought one sheep in and said he was going to shear it, and I said, "I'll shear it if you like," so he said, "Oh, can you shear?" I said I'd only done a few, so I started off shearing three thousand sheep on my own, that job. And I finished up and shore pretty near all of them apart from two or three days when another shearer came, and he was doing over three hundred each time.'

After one season of shed handing, Brian had a stand, working alongside Peter Spriggs. For the next few years, the two young lads developed their own local run around Aria, working the usual nine-hour shift, starting at 5 am. Norton Telfer later took Peter's place, and he and Brian worked together for two years. Romneys at the time were the predominant sheep, many of the woolsheds were just two or three stands and second shear was common, making for easier shearing. Their season started in September with hoggets and they would shear right through until Christmas, with second shear running February, March and April. Some farms they did eight-month fleeces, but generally they were shearing six-month fleeces.

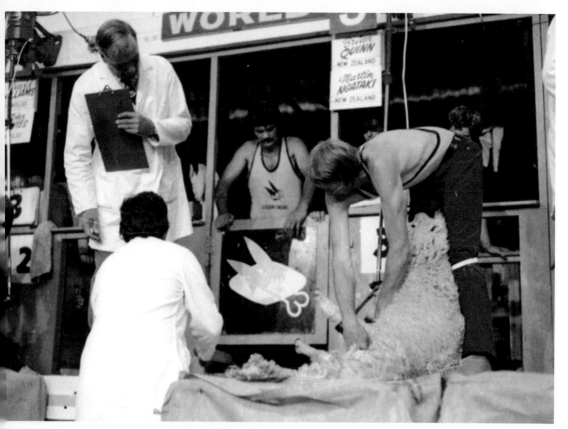

TOP Brian Quinn competes in the 1980 World Championship. Looking on is team mate Martin Ngataki. *Brian Quinn collection*

LEFT Poster boy Snow Quinn can still be found on woolshed walls today. Brewers' woolshed, Taranaki.

In that region at the time women were not working in the sheds, which tended to be open sheds with a lot of cockies and lads doing the work. They weren't churning out big numbers, shearing 400 a day between them.

Brian remembers that period fondly. Any shed over fifteen minutes away they would stay out at, living in the cocky's home, where they were well looked after. 'We used to work for Doreen and Abe Keely and we always stayed there. They had three girls and Doreen used to treat me like family; she was a lovely person.' He laughs at the memory of the three girls fighting over who would wash their shearing gear: 'Janice never had a look-in cos she was too small . . . it was lovely, absolutely lovely in those early days.'

Somewhere amidst all that shearing he happened to meet a young woman who was to become his wife. At the time, Lyn was teaching in Aria and boarding on a farm, and the couple met when Brian was shearing there; she brought beers down to the shed. Brian jokes, 'Only after meeting her I enjoyed going to school – spent more time there after hours than what I did when I was going to school.'

He talks of his marriage being a turning point in his life, with the decision to move to Alexandra. He quips that they 'had to make more money, so that's why we came down to the South Island'. They travelled in his Commer van, arriving 'down with all our worldly possessions in the back, Lyn on one side and a budgie in the middle and me driving'.

The decision to move was financially sound – and the weather was better, too.

'In those early days, we were getting three pound two-and-six, three pound five at home in the North Island, and we were getting six pound down here per hundred. We never had too many days off in a row, not like you did in the North Island. It was nothing to have ten days off in a row there in the North Island when it's wet, but down here if you have two or three days off you think your throat's cut.'

The season ran longer, too, so there was far greater security of work.

'I came down here end of April, early May, and we used to go and do a lot of crutching in May, June and July – be crutching full-time – then start shearing in the last week of July, and shear right through until January/February, and then start crutching again in March/April.'

Brian had set himself up with Murray McSkimming, the first of the contractors in Alexandra. At the time Murray employed all the top shearers and Brian got to work alongside some of them. They would come down for the pre-lamb shear in July through September.

'At that stage he probably had the best run down this way. He was very, very big when he started. He'd shear over a million; I'd imagine he was all over the place. He had all the best shearers in the country working for him . . . he had heaps, heaps of North Island boys used to come down – Ken Pike, Dawson Woodcock, Brian Thomas, Colin Bosher, Joe Ferguson, Bing Macdonald – all the top shearers came down and shore for him . . . Ray Alabaster . . . he had heaps of them.'

While talking of those gun shearers, Brian shared a story about Joe Ferguson that left a lasting impression on him:

'Another thing I admired with Joe: a lot of the time I'd be travelling, and he would be staying on the job and I mightn't get there till five to seven, and by the time you got your gear set up it was one minute past seven, but Joe never used to start until I was ready. And sometimes if I got there a bit later, me gear'd be set up . . . and all I needed to do was to put my mocs on and start. And he never started until I was ready; he just used to walk round and say, "There's no panic; just take your time, there's no hurry." I always admired him for that, but I learned a lot off him too.'

There were disadvantages as well to the South Island move; the colder climate was one:

'The temperature was a little bit against us for a start. I'd never been so cold in my life as when I went to Country Day at rugby down at Roxburgh; I didn't think it could possibly get that cold . . . so that took a little bit of adapting to. When I first came down I never had a heater in my car and after the first couple of trips nobody would ride with me. I'd spend most of the time scraping the ice off the windscreen.'

Another downside was the amount of travelling required:

'When I came down here I thought my throat was cut; we'd travel two hours to work. In saying that, when I first came down here in 1963 we only used to work eight hours – half past seven to half past five – but we did a lot more travelling.'

But there were benefits, too, as he remembers with warmth:

'My favourite stay-out place was up the lakes at Mount Nicholas. I did several years up at Mount Nicholas; Phil Hunt was the owner up there then and he used to really look after us. It was always a nice place to go. You used to go in by boat or fly – yip, occasionally we'd fly in if we were shearing at a competition somewhere and the boat had gone before we got back. We used to go and stay in there for ten days or two weeks. I only went up there in October for the wethers . . . we just used to sleep in the shepherds' quarters . . . some of the other boys used to go fishing straight after work. It was a lovely place to stay at and the tourists used to come in quite often on the boat and have a look around the shed, so that was always a bit more entertainment. They'd have them on a bit about different things . . . pretty all-near foreigners. That was before Queenstown was like it is now. Queenstown was good in the ski season, but it was a very one-horse town out of season: there was nothing there, all the fish-and-chip shops shut early, and if you were a bit late you couldn't get a feed . . . That's going back to the sixties.'

Some who worked with Murray never found him an easy man, but Brian built a good relationship with him and had a great deal of respect for him:

'I spent a lot of time with Murray in the early days, the early years; when he was contracting he'd be doing probably the work of four men. I don't know how he ever done it. He'd be out driving round all day and then all night driving round somewhere else – he was getting round the whole gang, and didn't employ anybody to help him, he'd be shifting them. It was just incredible, the amount of work he got through.'

Brian's quick to continue with a bit of a story about Murray, laughing as he tells it:

'He used to love shearing the stragglers, and he'd take guys up there and he'd shear two to their one all the time. The stragglers, they'd been out on the hill, some of them had two or three years' wool on them and they weren't very pleasant to shear; they weren't very nice. He used to love them. The smaller and harder they were, the better he loved them . . . He used to flail into them, get the wool off them. He used to love that. I'd say to him, "Gee, somebody's shearing well," and he said, "Oh, we'll see how well he does this week; I'm taking him up to shear stragglers." You'd always get the result, "How'd you go?" "No good." Especially the North Island boys. He used to love taking the North Island boys to shear a few stragglers. Some of the mobs, they'd have 150 to two hundred stragglers and they'd be every sort of sheep you could think of. Some would be hoggets, some would be wethers, some would be ewes, there'd be long-tailers . . . you name it, they'd have them. They must have gathered them all up at different stages and just kept them until they had enough to shear.'

The exact period he worked for Murray has been lost to memory, but Brian thinks it was under ten years. Wanting to stay closer to home, as all the travel was hard on his family, Brian changed contractors and went to work with Vic Harrex, who had mostly local sheds.

With only four or five shearers Vic was always able to shear himself, and it was a pattern that Brian himself followed later. The shift meant that over the summer months he worked for George Potae, as Vic didn't have much summer shearing. With the Potaes based in Milton, Brian still travelled, but worked it so that he stayed one night and travelled the next.

There was a change in culture and pace when he made the move.

'Vic was never a top shearer, but he was always a good clean shearer. He never shore big numbers of sheep, but he always shore clean; he had a pretty good reputation around the place; farmers liked him. It was a lot more laid-back. He didn't have the top shearers that Murray had working for him; it was more just going to work for the day and coming home again.'

Again, the flow of time has left exact dates a blur, but at some stage Brian became a contractor. It was not planned, but rather it was something that simply evolved.

As a contractor there was never any intention to grow too big; Brian preferred to stick with a team of five shearers as much as possible. 'If you get too top-heavy you can't shear yourself; things get out of hand, you're not there. I loved the shearing side of it, so that's why I kept shearing.' The run grew through Central Otago and Southland, along with work at Mataura over the summer, where his extended family would all come and work with him.

When it came time to pass on his business after running it for near-on thirty years, it was Dion Morrell who approached Brian about taking it over. Dion had first been in contact with Brian as a young up-and-coming

The end result of a lot of hard work. Tightly packed wool bales weighing in at somewhere between 180 and 200 kilos are ready to be loaded on the Muzzle Station flatdeck for the trip out to the woolstore.

shearer. He was living across the road and had popped over to see if Brian could give him a bit of help. Brian again sees the humour in the account:

I said, "How many you doing a run?" "Oh," he said, "most times a hundred, up round fifty an hour." I said, "Well I don't know if there's much I can show you – it's a long time since I done a hundred." But he was a really, really good shearer and a really nice guy.'

Though very much a family man, Brian was not opposed to a beer on the way home after work. Prior to going out on his own the gangs would call into possibly the Ōmakau, Poolburn, Ophir, Chatto Creek pubs, depending on where they were, for a drink.

'Just for half an hour; I was never one for having big sessions at the pub. Stop for half an hour and have a beer, then on your way. We always used to have a Sunday session, too, in Alex – at the Bendigo generally, or at the Middle Pub, the Criterion Club Hotel – and quite often if the police came round they'd knock on the door and you'd all run out and get in the cooler till they'd gone, and then go out again.'

The memory makes him laugh, as does the memory of Wattie Weir, the local constable, coming to drink with them sometimes:

'He'd be in there drinking with us on the Sunday session and then he had to go on duty, and he had to come and tell us to leave . . . But he was one of the standout policemen there as far as the shearers were concerned because he mixed with them and knew most of them. He used to hate telling everyone to leave, but it was part of his job.'

Reflecting back on his working life, Brian comments:

'I used to always try to do between 250 and 270 sheep on half-bred sheep, depending on the time of the year. And on the lambs, I liked to do over four hundred on the lambs and over three hundred on cross-bred ewes, but it didn't always work that way; it just depended on how the sheep went. You'd never go to work and say, "I'm going to hope to do 250 today," or, "I hope to do three hundred today." You'd just go and do what you could do. Over all the years I shore, I always just shore within myself. There was always the odd hard day here and there, but it wasn't a battle to see how many I could do every day.'

Like everything in life, though, there were exceptions:

'But any time I shore with a named shearer – you know, with a bit of speed about
him – I used to hit him from the first sheep; but there was no beg-your-pardon from
the first sheep on, and if they didn't let up it was an all-dayer. Go all day if I had to.
But quite often if you get four or five up for a start it takes all the pressure off, and
then you virtually shear along – you might shear sheep for sheep one run.'

There have been many highlights in Brian's shearing career, such as
travelling to the United Kingdom, and to Japan for the World Expo in 1970
to do shearing demonstrations. He has even shaken hands with the Queen.

Testament to Brian's humble nature is his ready acknowledgment
of all those who helped him on the path to success. And while he simply
states that shearing was 'the only job I've ever had – I just loved shearing',
he places it all in the context of his love for his family. 'I still maintain to
this day, if you haven't got family you haven't got much.'

The list of names of champion shearers and record holders is long, from
the great Raihania Rimitiriu in the early 1900s, to Johnny Hape, to the
Bowen brothers, Ivan and Godfrey, to Bing Macdonald, John Fagan,
Sir David Fagan, Alan MacDonald, Paul Grainger, George Potae, Colin
King, Eddy Reidy and Edsel Forde, through to today's crop of Johnny
Kirkpatrick, Rowland Smith and Nathan Stratford, never mind the line
of champion blade shearers, Bill Karaitiana, Donny Hammond, Peter
Burnett, Peter Casserly, and Tony Dobbs . . . then there are the women
who are starting to break records and make new ones: Jill Burney, Emily
Welch and Kerri-Jo Te Huia, and those from the past: Joy McCracken,
Maureen Hyatt, Margaret McAuley, Pam Warren and Ata Monds. There
are many more who have competed or attempted records whose names
have slipped through the net of fame, together with the legion of those
who have succeeded but are not mentioned here. All of them have had the
guts to go for glory.

Epilogue

It has been a privilege to delve into the world of shearing and to meet and interview shearers young and old who are passionate about an industry that is often underestimated and maligned. I wouldn't want to sanitise the industry, though; as Kelly Hokianga so eloquently philosophised, there are good shearers, there are fast shearers and there are arseholes. There's no sugar-coating the fact that the shearing industry is a gritty and tough working environment, and no place for those with delicate sensibilities. There are some problems with drugs and alcohol, but this is true in all walks of life. With increased health and safety demands placed on employers and a desire to push shearing's professionalism, contractors are openly grappling with the issues. When people are working in intense, stressful situations there is also the potential for cracks to appear in working relationships, causing blowouts and upset.

None of that diminishes the admiration that I feel towards those shearers, past and present, men and women, who have shorn and are shearing the millions of sheep that are bred in this country. We have shearers from around the world who come to learn and hone their skills here because New Zealand is home to some of the best shearers in the world. Our shearers have taken their skills and taught them overseas. The industry is rich with stories, humour and camaraderie. The sense of belonging that draws the shearing community together is one that creates a clear identity and, some would say, a sense of family.

For some, shearing will only ever be a job; but for many more in the industry there is something deeper that inspires them to 'keep walking in'. It has been described as a bug, a passion, an addiction; for some, especially those who have been exposed to work in the woolshed at an early age, it might just flow in the blood. As Nuku Hadfield relates:

'I do remember when our youngest son, Atawhai, was only two – he's always loved shearing – and he had a toy dog and he was in the kitchen here with me one day. He was making this noise – *mmmm* – and then I saw him throw the dog between his legs, and I thought, "Oh no, he's shorn his dog, and he's two years old."'

Farming contributors

I am indebted to the people listed below for agreeing to be interviewed for the initial work on this book. Their stories have enabled me to gain greater understanding of farming practices and regional variations that ultimately impact on the work of the shearer. And, of course, they introduced me to the shearers I went on to interview. Their interviews, along with those of the shearers, will be held at the Oral History Centre of the Alexander Turnbull Library, Wellington.

North Island –

Sam Jefferis – Waerenga, Waikato

Sam is the fourth generation of the Jefferis family to farm in the Waikato. Over the years the family have bought and sold a number of properties as their needs dictated. The original block taken up by Sam's great grandfather John Jefferis in 1876 on Jefferis Road is currently farmed by Sam's son Stuart. Sam is busy with the Stannard Road block bought in 1977. Combined the blocks make up 1300 acres. There is also a farm in Gisborne. Sheep and cattle have been the Jefferis' stock in trade although there was a period where dairying featured. Sam has moved from Romney to Romdales, a cross of Romney and Perendale. With their mix of land they are able to both breed and fatten. Sam shears to an eight-monthly cycle and with coarse wool prices low he sees shearing as simply about ensuring animal health.

Richard Brewer – Brewer Farms, South Taranaki

There have been Brewers farming in Taranaki for over 100 years. Richard and brother William are the third generation and they jointly run two Taranaki properties. With the mixed topography they are able to run 230 dairy cows, 2300 ewes and 360 cattle. A particular focus is fattening lambs for Coastal Spring Lamb, using a Poll Dorset terminal sire over Romney ewes. The lambs are a far cry from the little 'butter boxes' their father produced by crossing his Romney ewes with a Southdown ram. Rotational grazing of the sheep enables better feed management and has resulted in higher stocking rates. Richard's Bachelor of Agriculture from Massey has been a good foundation for his years farming.

John & Catherine Ford – Highlands Station, Rotorua

John and Catherine's 1240-hectare property just outside of Rotorua has been farmed by the Ford family since 1931. With much foresight Allen Ford, John's father, saw the potential in the land and pushed for it to be opened to farming. It became his life's work to develop the property. Ewe numbers peaked in the 1980s at 10,000. Since then the numbers have gradually decreased. In John and Catherine's time ewe numbers have come down to around 3200 as they have focused on yearling bulls, reflecting the dictates of the market. John is aware of the irony of having gone full circle with sheep breeds. In his father's day it was Romneys; John moved to composites, then Coopworths and now they have gone back to Romneys with half the ewes going to a Poll Dorset terminal sire. John and Catherine's profitable and sustainable farming practices were recognised when they were the recipients in 2015 of the Gordon Stephenson Trophy and National Farm Environment Award.

Geoff Candy – Marika Station, Rere, Gisborne

After growing up on his parents' properties around Gisborne, Geoff was well placed to take on his own farming career. As a child he was expected to help out around the farm and stock sense was ingrained in him at a young age. His father, at one stage a stock buyer in the region, may well have passed on his eye for stock to his son. Geoff shepherded for a number of years and worked as a fat-stock buyer for a time before purchasing his first property, Strathblane, which is now run by a manager. Geoff's Marika Station, where he currently lives, is a 3500-acre property carrying 500 cows and 3800 ewes. All the lambs are fattened, and Geoff and son Richard also buy upwards of 10,000 lambs a year and fatten them on another of their properties on the flats near Gisborne. Lambs can be fattened on Marika if there is excess feed.

Myles Mullooly – Whatatutu, Gisborne

Myles grew up on Journeys End Station, Matawai. His father Reg and mother Margaret were the second owners of the property. They purchased it from Cyril White, the original soldier settler. After twenty years of developing the property his parents sold it and bought Rahui Station, a 2000-acre block at Matawai. The timing of the shift in 1984, with Rogernomics and later the sharemarket crash, meant things were tough. Wh`ile they had Perendales up at Journeys End the easier contoured land of this next block at Matawai meant they went to a Romney flock. The family also leased a block at Motuhora and, between the properties, ran around 3000 ewes plus 200 beef cows and fattened bulls and steers. In 2006 Myles went farming in his own right: he bought the 315-hectare Whatatutu property and kept 340 hectares of the Matawai property. Between the properties he runs 2000 breeding ewes plus breeding cows. Replacement hoggets come down to the Whatatutu property, are put to the ram and go back to Matawai to lamb as two-tooths. The Whatatutu property has 500 ewes that go to a terminal sire, breeding cows and trading stock. Myles is the proud father of shearer Catherine Mullooly.

John Mitchell – Te Awapūtahi, Pōrangahau, Hawkes Bay

Tucked away in the hills of Pōrangahua is John's 640-hectare property that he has owned since 2004. The young farmer had always set his sights on farm ownership, having grown up on a farm in Gisborne and then Dannevirke. Before owning his own place John took on farm training at Telford training facility and then shepherding in both islands. When John took over Te Awapūtahi there were 3900–4000 ewes plus replacements and 110–120 cows. Facing three very dry years when he took over the property John found that the property was too highly stocked for the conditions he was facing so he has reduced ewe numbers to 3500 plus replacements of around 1000–1100 and the remaining lambs go to the works. Due to the nature of the land John carries 90% sheep and has approximately 70 cows. Hape Shearing came on board to shear when he first came on the farm and he moved to second shearing as he felt the animals fared better than when shorn just once a year.

Douglas Duncan – Otairi Station, Hunterville, Rangītikei

In 1881 John Duncan purchased the Otairi block of some 10,000 hectares. It was subsequently developed into an iconic sheep and beef station. Douglas, a great-great-grandson of John's, long held a dream of one day owning Otairi Station outright. That dream came true in 2013. He had grown up on family stories of Otairi Station but other than childhood visits he was not to experience station life there until his late teens when he worked as a shepherd there. Today Otairi is a 3300-hectare property, and is managed by Douglas's son Sam. The operation is focused largely on sheep breeding and finishing, as well as growing and finishing some 2300 steers. The station carries around 18,000 breeding ewes. Douglas's farms, Flockhart and Waipu, act as finishing units for Otairi lambs and cattle. Douglas has moved away from the traditional Romney to the Kelso, a composite breed with plenty of hybrid vigour.

Charlie Matthews – Waiorongomai, Wairarapa

Charlie is the sixth generation Matthews to run Waiorongomai. His great -great-great-grandparents purchased land in the Wairarapa in 1850. Like each generation before him Charlie has had the opportunity to try new and innovative things on the farm. While his father got rid of the Hereford stud, the pride of Charlie's grandfather, Charlie experimented with deer. Dairying and deer showed promise for a time but they eventually went by the wayside and beef and sheep have been the mainstay of the property. A Romney stud undergirded the property for generations but for financial reasons Charlie has recently let the stud go. He currently has 5500 ewes, replacement hoggets, and deals in trade lambs. Cattle have taken on greater significance on the property and he has introduced the Canadian Speckle Park beef breed to the mix. Charlie did a Diploma in Agriculture after leaving school, which laid the foundation for innovation. 'Wharepapa', the original property leased and then owned by Charles Matthews in 1850, has recently been repurchased by the Matthews family.

South Island –

Tina & Colin Nimmo – Muzzle Station, Clarence River, Kaikōura

The drive to own their own place saw pioneering farmers Tina and Colin take up land in one of the remotest locations in the country, tucked away behind the Seaward Kaikōura Range. Farming was nothing new to them: they had both grown up on farms and they had also worked together on Haldon Station in the Mackenzie Country. In 1980 they took over 18,000 hectares of pastoral lease from Bluff Station and named it Muzzle Station. There were no sheep on the station when they took up the lease, so they reintroduced merino to the area and focused on breeding from a Booroola ram gifted to them from Haldon to increase the flock's fecundity. Tina was quickly getting a healthy 100% lambing so they were able to build up their stock numbers. In 2007 the property went through tenure review and 10,000 hectares was taken over by DOC and the remaining 8000 hectares became freehold. They were also able to lease 8000ha of the Clarence Reserve. In 2015 Fiona and Guy Redfern, Tina and Colin's daughter and son-in-law, took over the station and it is currently running 2000 cattle and 4000 merinos.

Kate & Richard Foster – Terrace Station, Hororata, Canterbury

Kate grew up on Terrace Station. It was not until she was at secondary school that she discovered that her great-grandfather, Sir John Hall, was premier of New Zealand from 1879 to 1982. John Hall started leasing land in 1853 and built up a holding of 30,000 acres which was known as Rakaia Terrace Station. Before his death in 1907, he sold most of his land. The remaining land was divided between his two sons. Kate's grandfather took over 2600 acres, which became known as Terrace Station. It was later divided again, and Kate's father farmed over 1300 acres. Kate remembers helping in the shed during shearing as a young girl. When she was older she would use the windlass to haul bales of wool up to store them in the second storey of the woolshed.

Kate married Richard Foster in 1966 and they became staff on the station. A registered automotive mechanic by trade, Richard was not from a farming background but had always been keen to farm. At that time Kate's father was running 2000 Corriedale sheep and twenty beef cattle and 100 acres of crops were grown. There was a change to Romney and then Coopworths. Kate and Richard bought the property in 1971. They increased the sheep numbers to 3500, changed to Coopworths and improved the quality of the fleeces over a 35-year period. They weathered Rogernomics and the seven years of drought that struck at the same time during the 1980s. The Karaitianas were the regular shearers at the station for quite some years. Gradually the land size has been decreased and now nine hectares of the remaining land and buildings of this historic station are vested in the Terrace Station Charitable Trust.

Bruce Leadley – Wakanui, Mid Canterbury

Bruce, a retired school teacher, grew up on the 120-acre family farm in mid-Canterbury. While he never farmed himself, Bruce had memories of life on the farm and reflected back on the farming cycles of a small mixed farm typical of the Wakanui area east of Ashburton. Sheep numbers were anywhere from 150 to 200 and Bruce recalls driving them down to the neighbour's woolshed for shearing. Bruce's father had learned to shear at Ashburton Technical College at evening classes. Bruce also recalls working as a fleeco in other district woolsheds where professional shearers such as Hec Greene were employed.

Robyn & Ken Ferguson – Waipuna Station, Grey River Valley, West Coast

Ken is the third generation of Fergusons farming Waipuna Station. In his childhood his father was running 10,000 ewes plus some cattle on 5000 acres. Shearing time was exceptionally busy with upwards of 20,000 Romney sheep shorn, including lambs and hoggets. He has memories of shearer Malcolm Love, or 'Big Daddy' as he was known, and the Love gang working in the shed. They would bring sacks of kina with them and consume them for breakfast. Ken also recalls an oatmeal drink that they would make up in half-gallon bottles and drink in the shed. Seeing sheep as labour intensive, Ken over time has reduced sheep numbers down to 1300 ewes and 500 hoggets. He has developed a more mixed livestock model, stocking deer, cattle and sheep.

Robyn grew up on a cropping and sheep farm in Darfield, Canterbury. She married Ken in 1978 and shifted to Waipuna. She wryly comments that ever since, her role on the farm has centred on constantly feeding people, whether it be for docking, shearing or whenever people gather to work on the farm. Paying the shearers has been her domain as well. Over the years she has seen an increase in different nationalities of shearers – they even had a Mongolian shearer, the spelling of whose name caused consternation for IRD.

Robert Butson – Mount Nicholas Station, Queenstown

Robert Butson is a well-known woolgrower for the iconic merino clothing brand Icebreaker. In 1977 the Butson family took over the lease of the 40,000-hectare high-country property of Mount Nicholas on the shores of Lake Wakatipu. Robert's aim was to develop, through careful breeding, the flock of 10,000 half-breds to a flock of 20,000 22-micron merinos by introducing rams from Matangi Station in Alexandra. His father had advised him not to chase fashion but rather set his mind on something and stay his course and he would eventually reap the benefits; his father's advice has been proven true. Farming such a large high-country station is a far cry from his farming apprenticeship working the family farm in Garston with just 1000 sheep and 20–30 cattle with 40 acres of crops. Today the station is run by Robert's daughter and son-in-law, Kate and Jack Cocks.

Glossary

barrowing A learner shearer finishing the last side of a sheep for a shearer or shearing in their breaks.

blow Each stroke the shearer makes when shearing a sheep.

board The area where the shearers stand to shear the sheep.

bowyangs Leather strap or string tied around the shearer's leg below the knee to help keep their trousers from slipping down.

chip To tell off or chide a shearer if their work is not up to scratch.

closed board The design of a woolshed where the catching pen and porthole are opposite one another. The shearer drags the sheep from the pen across the board to the handpiece.

cockspur A piece of stainless steel attached to the back of the shears to prevent the blades from folding over to the wrong side.

cocky A farmer.

cotty A fleece when it is matted.

contractor The owner of a shearing business. They will often be running multiple gangs or teams of shearers and shed staff. They act as a middleman between farmer and shearer. They pay shearers and shed staff wages, provide transport to the sheds and some provide accommodation and meals.

cover comb/winter comb A specifically designed comb that leaves more of the fleece on the sheep to provide extra protection for the animal when shearing during colder months.

craic or crack Gaelic term roughly translated as fun or joy and love of life.

crawlies/yabbies Kōura or New Zealand freshwater crayfish.

cut-out A cut-out can refer to coming to the end of a particular mob or it can refer to finishing the entire shearing at a shed.

crutching Removing the wool around the tail and between the hind legs of a sheep.

dagging To remove the dags, or dried faeces, from the rear of a sheep.

ewe Female sheep.

expert A person who ground the gear and kept the shearing plant running. Each shearer now does their own gear.

fadge A wool pack.

fleeco The person picking up the fleeces (not commonly used today).

frib Wool from the brisket or chest of the sheep.

full wool When a sheep is shorn only once a year.

ganger A ganger manages the workers in the gang and is responsible for liaising with the farmer and the contractor.

Godfrey Bowen Godfrey and brother Ivan refined the shearing process into an efficient method of pattern shearing that became known as the Bowen technique, which revolutionised the industry. Godfrey was a key figure in the early Wool Board shearing schools and an ambassador for the shearing industry and is credited with increasing the profile and professionalism of the industry.

grinder The machine used to grind the shearer's cutters and combs.

gun An outstanding/expert fast shearer.

hogget A young male or female sheep at the stage where it is no longer considered a lamb but before a year old when its first two teeth come in and it becomes known as a two-tooth.

killers The sheep that farmers keep to kill for their own consumption.

main shear The pre-Christmas and early summer shearing is known as the main shear or the general shearing. All the flock will be shorn – hoggets, ewes and weaned lambs not yet ready for the works as well as rams. Second shear will happen some five months after that, depending on the region.

open shed Where a farmer employs, feeds and pays the shearer and shed staff directly. There is no contractor involved.

pizzle A sheep's penis.

porthole Where the sheep exit the shed into the counting pen.

pre-lamb shear The common practice of shearing ewes in winter just prior to lambing for animal health reasons and to prevent a wool break caused by the stress of lambing. Merino ewes are all shorn pre-lamb.

puha A green leafy vegetable, also known as sow thistle, used in the Māori boilup; often mistakenly called watercress.

raddle A coloured chalk/crayon used to mark a sheep that may have been cut.

ram Male sheep

ringer The fastest shearer in the shed; traditionally they took the No. 1 stand.

rousie/rouseabout Another term for woolhandler.

run A shearer's day is divided into runs – in an eight-hour day there are four two-hour runs with morning tea, lunch and afternoon breaks between each one. In a nine-hour day there is a two-hour run before breakfast then four one-and-three-quarter-hour runs. A run also refers to all the sheds that a contractor is contracted to work at throughout the year.

sheepo The individual who keeps the catching pens full of sheep. A shearer will call out 'sheepo' when they need more sheep in their pen.

shout To treat someone – usually in shearing terms a shout refers to providing the beer for everyone.

skirting Removing lower quality wool from the fleece.

smoko Morning and afternoon tea break.

tally The number of sheep a shearer shears in a day.

wether Castrated male sheep.

wheel Another name for a fast shearer.

woolhandler Member of the shearing gang who sweeps the board clear for the shearer, takes away the fleece and prepares the wool to be pressed.

wool classer A trained professional who classes or grades the wool according to such things as colour, length and micron or fibre thickness (shearers might teasingly call them 'the guesser').

Further reading

Books –

Bowen, Godfrey, *Wool Away! The technique and art of shearing*, Whitcombe & Tombs, Christchurch, 1955.

Brough, Tom, *The Way it Was: A farming, shearing, hunting life*, Fraser Books, Masterton, 2011.

Carter, Bill, & John MacGibbon, *Wool: A history of New Zealand's wool industry*, Wellington, Ngaio Press, 2003.

Laing, Doug, *Shear History: 50 years of Golden Shears in New Zealand*, Fraser Books, 2010.

Martin, John E, *The Forgotten Worker: The rural wage earner in nineteenth-century New Zealand*, Allen & Unwin/Trade Union History Project, Wellington, 1990.

Meadow, Graham, *Sheep Breeds of New Zealand*, Reed, Auckland, 1997.

Mills, AR, *Sheep-O! The story of the world's fastest shearers*, AH & AW Reed, Wellington, 1960.

Ogonowska-Coates, Halina, *Boards, blades & barebellies*, Benton Ross, Auckland, 1987.

Richards, Bill, *Off the Sheep's Back*, Benton-Guy, Auckland, 1991.

Riseborough, Hazel, *Shear Hard Work*, Auckland University Press, Auckland, 2010.

Williams, Des, *Top Class Wool Cutters: The world of shearers and shearing*, Shearing Heritage Publications, Hamilton, 1996.

Williams, Des & Margaret Way, *Last Side to Glory: The Golden Shears Open Championship 1961–1990*, Hazard Press, Christchurch, 1991.

Magazines –

Shearing: Promoting our industry, sport and people, Des Williams (ed.) Last Side Publishing, Hamilton, from 2002–

Shearing: The magazine for all in the shearing world, David Grace (ed.) Grace Editorial Ltd, Wellington, 1988–2000.

The New Zealand Shearer: The NZ Shearing Contractors' official magazine, Wellington, 1984–87.

Online –

Williams, Des, 'Shearing', *Te Ara – The Encyclopedia of New Zealand*, www.teara.govt.nz/en/shearing

Acknowledgements

While there is much that is solitary when writing, it most certainly does not happen in isolation. To that end there are many I must acknowledge and most sincerely thank.

First and foremost, I want to thank all my interviewees: the farmers and station owners, the shearers and shearing contractors. You all opened your homes, lives and woolsheds to this townie who wanted to understand your world. Without your willingness to be 'interrogated' there would be no book, so I am deeply indebted to you and am profoundly grateful. To the shearing gangs working in the sheds we visited, a sincere thank you for putting up with a keen photographer taking shots at every conceivable angle. I hope, if you find yourself in these pages, that we got your 'best' side.

To Fiona and Guy Redfern, Barry Pullin and Colin Nimmo, special thanks for making our trip to Muzzle Station possible. Suzanne Webby, a.k.a. the cook, thank you for taking us under your wing while we visited. Many thanks also to Sam Jefferis and Bart and Nuku Hadfield for specifically organising shearing in your sheds so Mark could take photos. To Brian Kerr, Bart and Tony Dobbs, thanks for the shearing lessons. Suffice to say I don't see a career change happening any time soon. To Des Williams, editor of *Shearing* magazine, a huge thank you for so willingly sharing your depth of knowledge about the industry and for providing me with past and current copies of the magazine.

The interviewing and travel could not have happened without my receiving generous funding through the New Zealand Oral History Award, administered through the Ministry for Culture and Heritage. Thank you to Alison Parr of the ministry for assisting me so invaluably in the development of the project. To Lynette Townsend, who took over Alison's role partway through the project, thank you for your support and sensitivity when the completion of the project was delayed.

To Lynette Shum and Linda Evans at the Oral History Unit of the Alexander Turnbull Library, thank you for your continued support and technical advice throughout this process. To the executive committee of NOHANZ, thank you for your support and encouragement and for your understanding when the writing deadline meant I missed meetings. To Lynley Simmons at the Timaru Library, thank you for always going the extra mile on my behalf. To my sister, Sarah Entwistle, who carried out the mammoth task of transcribing the interviews, sincere thanks, not just for the hours of typing but for your insight and gentle, gracious support.

It has been a real pleasure to have this book published by the team at Penguin Random House. To the ever-straight-shooting senior publisher Jeremy Sherlock, your expertise in drawing out the best in a writer makes it fun working with you. To project editor Louisa Kasza, thanks for keeping us all on task and grappling with the mountain of material being fired at you. To Megan van Staden, the wonderful designer, thank you for gathering the words and images into this fabulous book. The process of polishing the text was made easy by the deft hands of proofreaders Louise Belcher and Gillian Tewsley and editor Matt Turner. Thank you.

The bevy of friends and family around to support and encourage me throughout, particularly the writing phase, is always greatly appreciated. Special thanks to Adele, Chris, Helen, Linda, Janet, Kimble and Ailsa. Janet, thank you also for taking on some of the transcribing. To my brothers, Peter and John, thanks for chivvying me along when motivation was needed, and to my sister Catherine: flowers arriving on my doorstep, sent with love, just at a moment where encouragement was needed, spoke of your never-ending support and belief in me as a writer. Thank you. To my mum, Patricia Entwistle, you have taken on the task of being a very active parent-supporter without your loving offsider to share the load. Your care and love are so greatly appreciated. To my dad, Stewart Entwistle, for your interest and pleasure in seeing the birth of another book, and when you were physically no longer with us, for the sense of your presence with me as I wrote.

To my two precious cheerleaders, Charlotte and Kate, thank you for being there and putting up with an often distracted and possibly, though never to be confirmed, grumpy mother. Your support gives me strength. One of the absolute pleasures of working on this book has been travelling and working alongside my husband, Mark. His sensitivity and attention to the nuances of life in the shed have helped create photos that enhance the story of the shearers. Mark: for your wonderful photos, for your insight and the myriad of ways you support and encourage me – thank you.

Ngā manaakitanga,
Ruth Entwistle Low